CAMBRIDGE LIBRARY COLLECTION

Books of enduring scholarly value

Physical Sciences

From ancient times, humans have tried to understand the workings of the world around them. The roots of modern physical science go back to the very earliest mechanical devices such as levers and rollers, the mixing of paints and dyes, and the importance of the heavenly bodies in early religious observance and navigation. The physical sciences as we know them today began to emerge as independent academic subjects during the early modern period, in the work of Newton and other 'natural philosophers', and numerous sub-disciplines developed during the centuries that followed. This part of the Cambridge Library Collection is devoted to landmark publications in this area which will be of interest to historians of science concerned with individual scientists, particular discoveries, and advances in scientific method, or with the establishment and development of scientific institutions around the world.

Celestial Objects for Common Telescopes

Thomas William Webb (1807–85) was an Oxford-educated English clergyman whose deep interest in astronomy and accompanying field observations eventually led to the publication of his *Celestial Objects for Common Telescopes* in 1859. An attempt 'to furnish the possessors of ordinary telescopes with plain directions for their use, and a list of objects for their advantageous employment', the book was popular with amateur stargazers for many decades to follow. Underlying Webb's celestial field guide and directions on telescope use was a deep conviction that the heavens pointed observers 'to the most impressive thoughts of the littleness of man, and of the unspeakable greatness and glory of the Creator'. A classic and well-loved work by a passionate practitioner, the monograph remains an important landmark in the history of astronomy, as well as a tool for use by amateurs and professionals alike.

Celestial Objects for Common Telescopes

Thomas William Webb

CAMBRIDGE
UNIVERSITY PRESS

CAMBRIDGE UNIVERSITY PRESS

Cambridge, New York, Melbourne, Madrid, Cape Town, Singapore,
São Paolo, Delhi, Dubai, Tokyo

Published in the United States of America by Cambridge University Press, New York

www.cambridge.org
Information on this title: www.cambridge.org/9781108014076

© in this compilation Cambridge University Press 2010

This edition first published 1859
This digitally printed version 2010

ISBN 978-1-108-01407-6 Paperback

Celestial Objects for Common Telescopes

Thomas William Webb

CAMBRIDGE UNIVERSITY PRESS

Cambridge, New York, Melbourne, Madrid, Cape Town, Singapore,
São Paolo, Delhi, Dubai, Tokyo

Published in the United States of America by Cambridge University Press, New York

www.cambridge.org
Information on this title: www.cambridge.org/9781108014076

© in this compilation Cambridge University Press 2010

This edition first published 1859
This digitally printed version 2010

ISBN 978-1-108-01407-6 Paperback

CELESTIAL OBJECTS

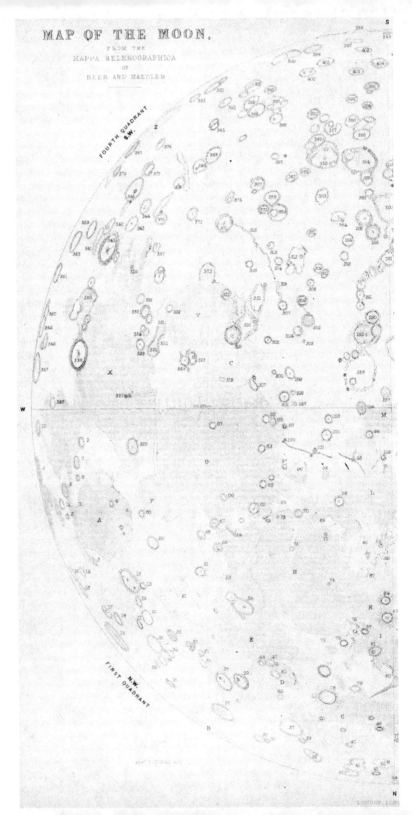

MAP OF THE MOON,

FROM THE

MAPPA SELENOGRAPHICA

OF

BEER AND MAEDLER.

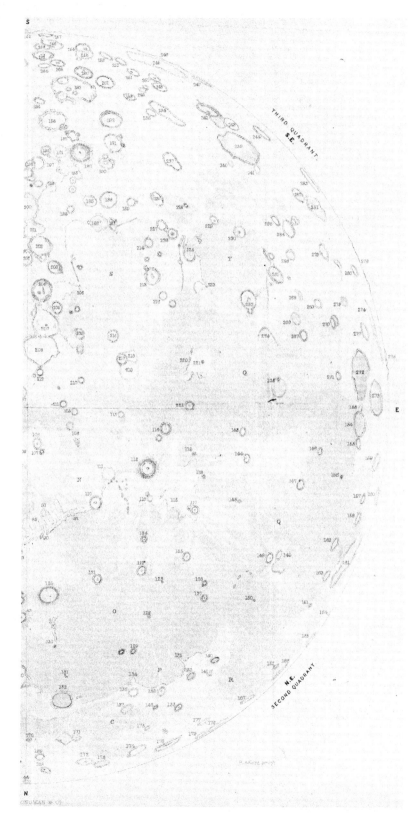

CELESTIAL OBJECTS

FOR COMMON TELESCOPES

BY THE REV. T. W. WEBB, M.A. F.R.A.S.

INCUMBENT OF HARDWICK, HEREFORDSHIRE

LONDON

LONGMAN, GREEN, LONGMAN, AND ROBERTS

1859

Many things, deemed invisible to secondary instruments, are plain enough to one who " knows how to see them." SMYTH.

When an object is once discovered by a superior power, an inferior one will suffice to see it afterwards. SIR W. HERSCHEL.

Inertia mors est philosophiæ — vivamus nos et exerceamur.
KEPLER.

TO

VICE-ADMIRAL W. H. SMYTH

K.S.F. D.C.L. F.R.S. ETC.

IN DUE ACKNOWLEDGMENT OF INDISPENSABLE ASSISTANCE

DERIVED FROM HIS MOST VALUABLE

CYCLE OF CELESTIAL OBJECTS

This little Work

IS RESPECTFULLY AND GRATEFULLY INSCRIBED

BY

THE AUTHOR.

Sic enim magnalia sapientiæ suæ decoravit Is, qui est ante sæculum et usque in sæculum : nihil redundat, nihil deficit, nec locus est censuræ cujusquam. Quam desiderabilia opera ejus! * * * * * et quis saturabitur videns gloriam eorum?

O qui lumine naturæ desiderium in nobis promoves luminis gratiæ, ut per id transferas nos in lumen gloriæ ; gratias ago Tibi, Creator, Domine, quia delectasti me in factura tua, et in operibus manuum tuarum exultavi.

KEPLER.

INTRODUCTION.

THE object of the following treatise is to furnish the possessors of ordinary telescopes with plain directions for their use, and a list of objects for their advantageous employment.

None but an eye-witness of the wonder and glory of the heavens can thoroughly understand how much they lose by description, or how inadequate an idea of them can be gathered in the usual mode, from books and lectures. It is but the narrative of the traveller instead of the direct impression of the scene. To do justice to this noble science,—to appreciate as we ought the magnificent testimony which it bears to the eternal Power and Godhead of Him "who by His excellent wisdom made the heavens," we must study it, as much as may be, not with the eyes of others, but with our own.

This, however, is no easy matter: nor is the want of a telescope the only difficulty. Instruments quite sufficient for the student's purpose are far less expensive than formerly; a trifling outlay will often procure them, of excellent quality, at second-hand; and many are only waiting to be called into action. But a serious obstacle remains to the inexperienced possessor. How is he to use his telescope in a really improving way? What is he to look for? And how is he to look for it? For want of an answer, many a good instrument is employed in a desultory and uninstructive manner, or consigned to dust and inactivity.

Materials for his guidance exist, indeed, in profusion; but some of them are difficult of access; some, not easy of interpretation; some, fragmentary and incomplete: and the student would find it a discouraging task to reduce them into a serviceable form. This, then, is what has been attempted for him in the following pages, by one who, during many years, would have rejoiced to avail himself of some such assistance, if he had known where to meet with it, and who does not know where it is to be met with, in a convenient shape, to the present day.

For the more advanced observer, the "Cycle of Celestial Objects," published in 1844 by Captain, now Vice-Admiral Smyth, will be found a treasury of varied information, and of the highest value as the companion of a first-rate telescope: but its very superiority, to say nothing of its bulk and cost, renders it more suitable for his purpose, than for those humble beginnings which are now in view. It has, however, been of the most essential service in the preparation of the present undertaking, which without it would, in all probability, never have seen the light, and which, as far as the sidereal portion of it is concerned, is based upon it as the standard authority.

Nothing would have been easier than, on so fertile a subject, to have expanded this treatise to a much larger bulk: but it would thus, in some measure, have defeated its own object. In order therefore to reduce the size of the volume, without omitting such details as may seem to be required by the present state of Astronomy, the reader will have to excuse a condensed mode of expression, the result of necessity rather than of choice; and, as considerable pains have been taken in the verification of facts, a general list of authorities will supersede references at the foot of the page.

Limited in extent, imperfect in execution, and in parts only suggestive in character, this little book may perhaps serve as a foundation, on which students of astronomy may raise the superstructure of their own experience; and in that case the author's intention will be fulfilled. He will be especially gratified, if his endeavour to remove some difficulties may tend to increase

the number of those who " consider the heavens." For
he is convinced that in such a personal examination of
their wonders will be found an interesting and delightful
pursuit, diversifying agreeably and instructively the
leisure hour, and capable of a truly valuable application,
as leading to the most impressive thoughts of the littleness of man, and of the unspeakable greatness and glory
of the CREATOR. To such a study, the impressive words
of the late Sir R. H. Inglis may be most suitably applied : " Every advance in our knowledge of the natural
world will, if rightly directed by the spirit of true hu-
mility, and with a prayer for GOD's blessing, advance us
in our knowledge of Himself, and will prepare us to
receive His revelation of His Will with profounder
reverence."*

* Report of British Association, 1847.

CONTENTS.

CONTENTS.

AUTHORITIES.

Annals of Harvard College Observatory.
Annuaire du Bureau des Longitudes.
Astronomische Jahrbuch.
———————— Nachrichten.
Beer and Mädler, Der Mond.
———————— Beiträge, &c.
Bianchini, Hesperi et Phosphori, &c.
Bond, Account of Donati's Comet.
Breen, Planetary Worlds.
Comptes Rendus.
Donati, Essay on Comets.
Gould (American), Astronomical Journal.
Gruithuisen, Naturgeschichte, &c.
———————— Neue Analekten.
Herschel, Observations at Cape of Good Hope.
———————— Outlines of Astronomy.
Hind, Solar System.
——— The Comets.
Histoire de l'Académie.

Hooke, Posthumous Works.
Humboldt, Cosmos.
Huygens, Systema Saturnium.
Kepler, Opera.
Lohrmann, Topographie, &c.
Memoirs of Royal Astronomical Society.
———————— American Academy.
Memorie del Collegio Romano.
Monthly Notices of R. Astron. Society.
Philosophical Transactions.
Reports of British Association.
Schmidt, Resultate, &c.
Schröter, Selenotopographische Fragmente.
——— Aphroditographische ditto.
——— Beobachtungen über den Cometen von 1807.
——— Ditto, 1811.
Smyth, Cycle of Celestial Objects.
——— Ædes Hartwellianæ.
Struve, Mensuræ Micrometricæ.

Pulchra sunt omnia, faciente Te, et ecce Tu inenarrabiliter pulchrior, qui fecisti omnia.

ST. AUGUSTINE.

CELESTIAL OBJECTS.

PART I.

THE INSTRUMENT AND THE OBSERVER.

O multiscium et quovis sceptro pretiosius Perspicillum! an, qui te dextra
tenet, ille non rex, non dominus constituatur operum Dei? Vere tu
  Quod supra caput est, magnos cum motibus orbes
  Subjicis ingenio.— KEPLER.

THE TELESCOPE.

ALTHOUGH the professed design of this volume is to provide a
list of objects for common telescopes, it may not be out of
place to premise a few remarks upon the instruments so
designated.

By "common telescopes" are here intended such as are
most frequently met with in private hands; achromatics of
various lengths up to 5 or $5\frac{1}{2}$ feet, with apertures* up to $3\frac{3}{4}$
inches; or reflectors of somewhat larger diameter, but in
consequence of the loss of light in reflection, not greater
brightness.† The original observations in the following pages

---

* "Aperture" always means the clear space which receives the light
of the object; the diameter of the object-glass in achromatics, or the
large speculum in reflectors, exclusive of its setting.

† Maskelyne estimated the apertures of reflectors and achromatics of

were almost entirely made with such an instrument, an achromatic by the younger Tulley, $5\frac{1}{2}$ feet in focal length,* with
an aperture of $3\frac{7}{10}$ inches, and of fair defining power; smaller
instruments of course will do less, especially with faint
objects, but are often very perfect and distinct: and even
diminutive glasses, if good, are not to be despised; they will
shew *something* never seen without them. I have a little
hand telescope, $22\frac{1}{4}$ inches long when fully drawn out, with
an object-glass of about 14 inches focus, and $1\frac{1}{3}$ inch aperture: this, with an astronomical eye-piece, will shew the
existence of the solar spots, the mountains in the Moon,
Jupiter's satellites, and Saturn's ring. Achromatics of 4
inches and upwards are becoming much less expensive, and
will soon be more common; even for these it is hoped that
this treatise, embodying the results of the finest instruments,
may not be found an inadequate companion as far as it goes.

In buying a telescope, we must disregard appearances.
Inferior articles may be showily got up, and the outside must
go for nothing. Nor is the clearness of the glass, or the
polish of the mirror, any sign of excellence: these may exist
with bad " figure " (*i.e.*, irregular curvature), or bad combination of curves, and the inevitable consequence, bad performance. Never mind bubbles, sand-holes, scratches, in object-

equal brightness as 8 to 5. Dawes gives this value for Gregorians, but
like Herschel II. rates Newtonians as 7 to 5. Steinheil has recently
ascribed much more light to achromatics. Arago strangely supposed
none was lost in them. Secchi thinks the Roman achromatic,
$9\frac{6}{10}$ inches aperture, equal to Herschel II.'s reflector of $18\frac{1}{2}$ inches.

* The focal length is measured from the object-glass, or speculum, to
the spot where the rays cross and form a picture of the sun or any celestial body.

glass or speculum; they merely obstruct a very little light. Actual performance is the only adequate test. The image should be neat and well defined with the highest power, and should come in and out of focus sharply; that is, become indistinct by a very slight motion on either side of it. A proper test-object must be chosen; the Moon is too easy; Venus too severe except for first-rate glasses; large stars have too much glare; Jupiter or Saturn are far better; a close double star is best of all for an experienced eye; but for general purposes a moderate sized star will suffice; its image, in focus, with the highest power, should be a very small disc, almost a point, accurately round, without "wings," or rays, or mistiness, or false images, or appendages, except one or two narrow rings of light, regularly circular, and concentric with the image;* and in an uniformly dark field; a slight displacement of the focus either way should enlarge the disc into a luminous circle. If this circle is irregular in outline, or much better defined on one side of the focus than the other, the telescope may be serviceable, but is not of much excellence. The chances are many, however, against any given night being fine enough for such a purpose, and a fair judgment

---

* There is something unexplained about these rings. They are commonly ascribed to the diffraction or inflection of the light which grazes the brass setting of the object-glass: yet they are not, as might be expected, always the same with the same instrument. Herschel II. speaks of nights of extraordinary distinctness, in which "*the rings are but traces of rings,* all their light being absorbed into the discs." I have entered 1852, March 23, as "a very fine night, though the rings and appendages around the brighter stars were rather troublesome;" 1852, April 1, "an exceedingly fine night at first, with scarcely a trace of rings or appendages." See also the star 70 Ophiuchi, in the following catalogue.

may be made by day from the figures on a watch-face, or a
minute white circle on a black ground, placed as far off as is
convenient.   An achromatic, notwithstanding the derivation
of its name, will shew colour under high powers where there
is a great contrast of light and darkness.   This " outstanding "
or uncorrected colour results from the want of a perfect
balance between the optical properties of the two kinds of
glass of which the object-glass is constructed : it cannot be
remedied, but it ought not to be obtrusive.   In the best
instruments it forms a fringe of violet, purple, or blue, round
luminous objects in focus under high powers, especially Venus
in a dark sky.   A red or yellow border would be bad ; but
before condemning an instrument from such a cause, several
eye-pieces should be tried, as the fault might lie there, and be
easily and cheaply remedied.   Reflectors are free from this
defect ; but can hardly be recommended to beginners.   They
are noble instruments when well made, and the focal picture
is very beautiful ; but the mirrors are liable to tarnish, or get
out of adjustment, excepting in experienced hands, and for
this reason, as well as from its greater frequency, the achromatic will be the chief, though not the exclusive subject of
our remarks.

The eye-piece is only a kind of microscope, magnifying the
image formed in the focus of the object-glass or speculum.
The size of this image being in proportion to its distance from
the glass or mirror which forms it, the power of the same eye-
piece in different telescopes varies with the focal length.
Hence one disadvantage of a short telescope ; to get high
powers, we must employ deeply-curved and minute lenses,
which are much less pleasant in use : with a telescope twice
as long, half the power in the eye-piece produces an equal

effect. The focal picture, as in the camera, is always inverted, and so in the astronomical eye-piece it remains.* For terrestrial purposes it is erected by two additional lenses; but a loss of light is thus incurred, and as the inversion of celestial objects is unimportant, erecting eye-pieces (always the longest of a set) should never be employed for astronomy; the eye soon becomes accustomed to the inverted picture, and the hand to the reversed motion in following the object. A multitude of eye-pieces is needless, but three at least are requisite; one with low power and large field, for extended groups of stars, nebulæ, and comets, supplying also, if necessary, the place of a "finder" for deeper magnifiers; a stronger one for general purposes, especially the moon and planets; and a third, as powerful as the telescope will bear, for minuter objects, especially double stars. A greater number of eye-pieces admits, however, of what is often important, an adaptation of the power to the brightness of the object. Ordinary astronomical eye-pieces are shorter in proportion to their power. It is a better plan to have them to slide into a tube than to adapt them by a screw; in which case they are more liable to be dropped and injured. The power may be much increased by unscrewing and taking away the "field-lens,"—that furthest from the eye; but the field will be less perfect, excepting near its centre. The highest powers of large telescopes are often made thus, with single lenses, but the convex face of the lens is then turned towards the eye, as it gives sharper vision. A concave field-lens increases the power materially, with, at

* It is erect in the Galilean eye-piece and the Gregorian reflector. But the use of the former is almost confined to opera-glasses, as its field with high powers is exceedingly small; and the latter is an inferior construction, and now little adopted.

the same time, shallower curves, and consequently less ab-
erration and more ease to the eye; its power, too, may be
varied by varying the distance between the lenses; but it is
not usually made or sold. The common kind, with two lenses,
having the flat side of each next the eye, is called the Huygenian, achromatic, or negative eye-piece.

The amount of power in valuable instruments is marked on
the eye-piece by the maker.* If this has not been done, we
may approximate to it thus: unscrew the lens next the eye;
hold it so that the sun may shine straight through it; bring a
little rod or pin of wood or bone, which will not scratch, carrying a sliding bit of card, to touch the back (or flat side) of the
lens; slide the card till the image of the sun upon it is as small
and distinct as possible; measure with compasses the distance
of the card from the end of the rod, and you have the focal
length of the lens; divide by this the focal length of the object-
glass, † found by the distance at which it gives a perfect image
of the sun or moon, on transparent paper stretched across the
small tube from which the eye-piece has been removed, — the
quotient is the power of that lens used alone. Do the same by
the other — the "field-lens;" this cannot quite so conveniently
be used alone, but from the two foci we may compute the
power of a Huygenian eye-piece, by the following rule:
*Divide twice the product of the focal lengths of the two
lenses, by the sum of their focal lengths; the quotient is the
focal length of an equivalent single lens.* In general, the foci

* Not, however, always so as to be depended upon. The celebrated
Short exaggerated the powers of his reflectors; and those of the great
achromatics of Dorpat and Berlin were found by Struve and Encke to
be overrated.

† Or large speculum of a *Newtonian* reflector.

of the lenses are as 3 to 1; this gives $1\frac{1}{2}$ for the combined
effect; or the power of such an eye-piece is usually equal to
two thirds the power of the eye-lens used alone. The measurement of small lenses is a delicate process; but if, with the
measure in the compasses, we *step* carefully along a considerable length of a scale of equal parts, and divide the result by
the number of steps, we shall increase its correctness; and if
we measure afresh, and repeat the stepping process several
times, and take a final average, it will not be far from the truth.

A good stand is essential: if unsteady, it will spoil the most
distinct performance; if awkward, it will annoy the observer;
if limited in range, it may disappoint him at some interesting
juncture. It may be well left to a respectable optician; but
where expense is a serious consideration, a little mechanical
ingenuity and knowledge of such contrivances will devise one
which will answer sufficiently. The old arrangement, with a
vertical and horizontal, or " altitude and azimuth " motion, is
simple and manageable: the equatorial form, which makes the
telescope revolve on an axis parallel to that of the earth, has
some great advantages, in following the object by a single
motion, and where the expense of divided circles and spirit-
levels is admissible, in finding planets and bright stars by day,
and identifying minute objects by night: but, to do its work,
it must be placed accurately in the meridian, and out of that
position has little advantage. In any case, if the stand is to
be moveable, let it be strong enough for steadiness without
being too heavy for portability.

An object-glass of inferior defining power may sometimes be
improved by stopping out defects, or contracting the aperture.
Streaks or specks of unequal density are very injurious: they
may be detected by turning the telescope to a bright light,

taking out the eye-piece, and placing the eye in the focus; every irregularity will then be visible in the illumination which overspreads the object-glass; and, if of small extent, may be stopped out by a bit of sticking-plaster. If the telescope is not thus improved, try a contracted aperture : make a cap of pasteboard fitting over the object-glass like the usual brass cap, but with a circular opening a little less than the clear aperture ; —if the indistinctness is thus diminished but not removed, try several discs of pasteboard placed successively within this cap, with progressively contracted openings, till distinct vision is obtained; there we must stop, or valuable light will be lost. An excentric opening in the pasteboard cap may sometimes be serviceable, being turned round the axis so as to conceal different parts of the glass or mirror, till the best effect is produced : in other cases, a central pasteboard disc, supported by narrow arms from the sides, and leaving an open ring of light all round, may be tried. But, for comets or nebulæ, it will be best to restore the original aperture, as with faint and ill-defined objects light is more essential than distinctness. Mr. Grove has found that the insertion of a small lens, (not so small as to waste light) will sometimes produce a surprising improvement in an ordinary achromatic; it is to be of plate-glass, plano-convex, the plane side next the eye, of rather longer focus than the object-glass, and placed behind it, at $\frac{1}{4}$ or $\frac{1}{5}$ of the focus of the latter; the exact position to be ascertained by trial. To see whether the smaller speculum which stands in the mouth of a reflector is rightly centered, unscrew the eye-piece, and look at the small speculum through the opening so made; in it will be seen an image of the large mirror, with the small speculum reflected again upon it—these images should be concentric with the small speculum; if not, a very cautious alternate loosening and tightening of the three little screws in a triangle at the back of the small speculum will bring all right.

We will close this section with the encouraging words of the Council of the Royal Astronomical Society, in their Report for 1828. "Every one who possesses an instrument, whose claims rise even not above a humble mediocrity, has it in his power to chalk out for himself a useful and honourable line of occupation for leisure hours, in which his labour shall be really valuable, if duly registered; . . . those who possess *good* instruments, have a field absolutely boundless for their exertions."

THE MODE OF OBSERVATION.

AN ordinary telescope may be easily prepared for use: to fix it on its stand; to point it by means of the finder; to adjust the focus to the eye (remembering that different eyes require different adjustments), are processes scarcely requiring instruction. But many mistakes may be made in detail; and in this, as in everything else, there are various methods of doing the thing the wrong way. The present section will, therefore, consist of negative rather than positive directions, pointing out rather what should be avoided than what should be done.

1. Do not begin by fixing the telescope in a warm room, and opening the window. A boarded floor is bad, as every movement of the observer is liable to produce a tremor; but the mixture of warm and cool currents at the window is worse; it is an artificial production of the fluttering and wavering

which, as naturally formed, are such an annoyance to astronomers. If a window must be used, let it be opened as long
beforehand as may be, and let the object-glass be pushed as
far as possible outside; the nuisance may thus be sometimes
abated, but the right place is unquestionably out of doors.

2. Do not wipe the object-glass or speculum more than
can possibly be helped. Hard as the materials are, scratching
is a very easy process; and the result of ordinary wiping
may be seen in an old spectacle-glass held in the sunshine.
The most valuable part of a good telescope deserves much
more careful treatment; and, if protected from dust and damp,
it will very seldom require to be touched. Nothing but great
carelessness would expose it to dust; and the dewing of the
surface may be almost always avoided. The object-glass or
speculum, if kept in a cold place, should not be uncovered,
if possible, in a warmer air till it has gained something of its
temperature; and it must be invariably closed up in the air
in which it has been used; or, in either case, it may be
dewed like a glass of cold water brought into a heated room.
The object-glass, however, being so much exposed to radiation, requires additional protection; and this may be easily
contrived. A tube of pasteboard, or tin, or very thin wood,
such as is used for hat-boxes, fitting on to the place whence
the brass cap has been removed, and three or four times
longer than wide, will, in general, keep the object-glass bright.
This " dew-cap" must fit tight enough to stand firm, or it will
bend down and intercept the light; but not so tight as to
cause trouble in removing it to put on the brass cap in the open.
air. It is better to blacken its interior—indeed, necessary, if of
tin; this may be done with lamp-black mixed with size or
varnish, so as neither to shew a gloss nor rub off; or a piece

of black cloth or velvet may be glued or pasted inside
it. A small dew-cap on the finder will often save much
trouble. Should it be necessary to leave the telescope for
some time in the air, a clean handkerchief thrown over the
end of the dew-cap will be an additional safeguard. Should
an object-glass or speculum become damped after all, do not
close it up in that state; if the cloud of dew is very slight,
it may quite disappear in a warm room; if dense, however,
it may leave a stain which ought to be quickly removed, as
well as any little specks of dirt or dulness which will form,
one knows not how. To do this, dust the surface first with
a soft camel's hair pencil or varnishing brush, which will
remove loose particles; then use, very cautiously, a very soft
and even piece of chamois leather, which has not been employed for any other purpose, and must be always kept
wrapped up from dust; or a very soft silk handkerchief (which
Lassell uses for glass) preserved with similar care. But the
wiping must be as gentle as possible; hard rubbing would
certainly damage a speculum, and do no good to an object-
glass. Any refractory stains may be breathed upon, or touched
with rectified spirits of wine, and wiped till dry. A slight
tarnish may frequently be removed from a speculum by
lemon-juice, or a solution of citric acid, carefully wiped off
in a short time; if this does not restore its brightness, it is
better to leave it alone; a slight loss of light is not so great
an injury as would result from strong friction. The taking
out or replacing of an object-glass or mirror is a delicate
operation, and hurry or carelessness may easily make it a
very dangerous one; speculum metal is nearly as brittle as
glass.

To clean eye-pieces, unscrew the lenses and wipe them as

above; a small pointed bit of soft wood may be required to press the leather to the edges of the glass. Lenses too small to be got at conveniently must be cleaned with a camel's hair pencil.

Look occasionally at the brass-work, and rub it with leather, but use no polishing powder, which might injure the lacquering.

3. If the telescope does not seem altogether right, notwithstanding all the pains you can take in bringing it to focus, do not meddle with screws or adjustments, unless you thoroughly understand the construction, or can obtain good directions. You may centre the small speculum of a reflector with safety, but in most cases a screw-driver is a dangerous tool in inexperienced hands.

4. Do not use any part of a telescope or stand roughly, or expose it to any blow or strain. It is a delicate instrument, and well deserves careful preservation.

5. Do not spare trouble in adjusting the focus. It is well known that it must be altered for the different distances of objects: but it would hardly have been supposed that rays proceeding from a body a quarter of a million of miles off would require a different adjustment from those sensibly parallel: yet such is the case, the focus for the moon being perceptibly longer than for the stars: different eyes also require a change, sometimes a great one; and the same observer's focus is not invariable, being affected by the state of the eye and the temperature of the tube.

6. Do not over-press magnifying power. Schröter long ago warned observers against this natural practice, which is likely to lead beginners into mistakes. A certain proportion of *light* to *size* in the image is essential to distinctness; and though by

using a deeper eye-piece we can readily enlarge the size, we
cannot increase the light so long as the aperture is unchanged;
while by higher magnifying we make the inevitable imperfection of the telescope more visible. Hence the picture
becomes dim and indistinct, beyond a certain limit, varying
with the brightness of the object, and the goodness of the
telescope. Comets and nebulæ, generally speaking, will bear
but little magnifying. For the moon and planets, the power
should be high enough (if the weather will bear it) to take off
the glare, low enough to preserve sufficient brightness and
sharpness: the latter condition being preserved, minute details
are likely to come out better with an increase of power. Stars
bear much more magnifying, from their intrinsic brilliancy;
and they are enlarged very slightly in proportion: their images,
which would be absolute points in a *perfect* telescope, ought
never, ·with any power, to exceed the dimensions of minute
discs, — *spurious discs*, as they are termed, arising solely from
optical imperfection, and smallest in the best telescopes. A
very high power has, however, so many disadvantages, in the
difficulty of finding and keeping the object, the smallness of
the field, the rapid motion of the image (in reality, the magnified motion of the earth), and the exaggeration of every
defect in the telescope, the stand, and the atmosphere, that the
student will soon learn to reserve it for special objects and for
the finest weather, when it will sometimes tell admirably. A
very low power is apt to surround bright objects with irradiation, or glare. Experience in all these matters is the surest
guide.

7. Do not be dissatisfied with first impressions. When
people have been told that a telescope magnifies 200 or 300
times, they are often disappointed at not seeing the object

larger. In viewing Jupiter in opposition with a power of only
100, they will not believe that he appears between two and
three times as large as the moon to the naked eye; yet it is
demonstrably so. There may be various causes for this illusion;
— want of practice, — of *sky-room*, so to speak, — of a standard
of comparison. A similar disappointment is frequently felt in
the first impression of very large buildings; St. Peter's at
Rome is a well-known instance. If an obstinate doubt remains,
it may be dissipated for ever when a large planet is near
enough to the moon to admit of both being seen at once, the
planet through the telescope, the moon with the naked eye.

8. Do not lose time in looking for objects under unfavourable circumstances. A very brilliant night is often worthless
for planets or double stars, from its blurred or tremulous
definition; it may serve, however, for nebulæ, which have no
outlines to be deranged : a hazy or foggy night will blot out
nebulæ and minute stars, but sometimes defines bright objects
admirably; never condemn such a night untried. Look for
nothing near the horizon; unless, indeed, it never rises much
above it; nor over, or to the leeward of a chimney *in use*,
unless you wish to study the effect of a current of heated air.
If you catch a really favourable night, with sharp and steady
vision, make the most of it : you will not find too many of
them. Smyth, who thinks our climate has been unfairly de-
preciated, says, " where a person will look out for opportunities
in the mornings as well as evenings, and especially between
midnight and daybreak, he will find that nearly half the nights
in the year may be observed in, and of these sixty or seventy
may be expected to be splendid." But ordinary students must
of course take their chance, with their fewer opportunities.
With due precaution, nothing need be feared from "night-air: "

that prejudice is fully confuted by the well-known longevity of astronomers, even of such as have habitually protracted their watchings

" Till the dappled dawn doth rise."

9. In examining faint objects, do not prepare the eye for seeing nothing, by dazzling it immediately beforehand with a lamp, or white paper. Give it a little previous rest in the dark, if you wish it to do its best.*

10. When a very minute star or faint nebula is not to be seen at once, do not give it up without trying *oblique* or *averted vision*, turning the eye towards the edge of the field, but keeping the attention fixed on the centre, where the object ought to appear; this device, with which astronomers are familiar, is often successful ; its principle depends probably on the greater sensitiveness of the sides of the retina.

11. Do not avoid the trouble of recording regularly all you see, under the impression that it is of no use. If it has no other good effect, it tends to a valuable habit of accuracy: and you might find it of unexpected importance. And, like old Schröter, *trust nothing to memory.* If there has been haste, — and sometimes if there has not, — it is surprising what unforeseen doubts may arise the next day: make at least rough notes at the time, and reduce them speedily into form, before you forget their meaning.

12. Do not be discouraged by ignorance of drawing from attempting to represent what you see. Everybody ought to be able to draw; it is the education of the eye, and greatly

---

* Herschel II., when about to verify his father's observations on the satellites of Uranus, prepared his eye with excellent effect, by keeping it in utter darkness for a quarter of an hour.

increases its capacity and correctness : but even a rough sketch may have its use; taken on the spot, it will not be all untrue; it may secure something worth preserving, and lead to further improvement.

In conclusion, may I be permitted to remind the young observer, not to lose sight of the immediate relation between the wonderful and beautiful scenes which will be opened to his gaze, and the great Author of their existence? In looking upon a splendid painting, we naturally refer its excellence to the talent of the artist; in admiring an ingenious piece of mechanism, we cannot think of it as separate from the resources and skill of its designer; still less should we disconnect these magnificent and perfect creations, so far transcending every imaginable work of art, from the remembrance of the Wisdom which devised them, and the Power which called them into being. Such is eminently the right use of the Telescope, —as an instrument, not of mere amusement or curiosity, but of a more extensive knowledge of the works of the Almighty. So new an aspect as has thus been given to the material universe, — so amazing a disclosure as has thus been permitted to man, of the vastness of his Maker's dominion,— can hardly be ascribed to blind accident or human contrivance : in thus employing Galileo's invention, we may well feel his grateful acknowledgement, that it was the result of the " previous illumination of the Divine favour," * to have been not only beautiful, but true.

* Divina prius illuminante gratia.

PART II.

THE SOLAR SYSTEM.

O domus luminosa et speciosa, dilexi decorem tuum, et locum habitationis
gloriæ Domini mei, fabricatoris et possessoris tui!—St. Augustine.

THE SUN.

The solar phenomena are especially wonderful. The unrivalled pre-eminence of that glorious sphere,—the dependence
of our whole system upon the mysterious processes developed
at its surface,—the rapid and extensive disturbances in their
action, as well as the daily visibility of the object, all combine
to invite research. But the student had better not begin
here : more than one astronomer has suffered from that
piercing blaze : Galileo probably thus blinded himself
wholly, and Herschel I. in part. With due precaution, there
is no danger; but the eye and hand had better first acquire
experience elsewhere. Much depends on the dark glass of the
solar cap which is to be screwed on the eye-piece; red is
often used, and *may* be dark enough—it is not so always—
but it transmits too much heat; green is cooler, but seldom
sufficiently deep. The Germans have employed deep yellow,
probably to save the brightest and most central rays of the
spectrum. Herschel I. adopted, with great success, a trough

C

containing a filtered mixture of ink and water. Cooper (of
Markree Castle, Ireland) uses a "drum" of alum water and
dark spectacles, and can thus endure the whole aperture, $13\frac{3}{10}$
inches, of his great 25 feet achromatic. Mr. Reade has found
common eye-pieces improved by filling the interior with
water; and this plan might be applied to the Sun, the water
being darkened with ink; but probably the detained heat
might cause some troublesome result in a long observation.
He has also stated that where screen-glasses are used, a small
eye-hole between the eye-piece and the solar cap will save
the glass from being cracked by the heat, which often occurs
with apertures of more than 3 or 4 inches.* Combinations
of colour are good. The Americans speak well of red and
green. Herschel II. uses green and cobalt blue.† If there is
to be only one solar cap, deep bluish grey, or neutral tint,
will be quite satisfactory; if several, it would be worth while
to have different colours, Secchi's observations at Rome seeming to shew that the visibility of very delicate details may
depend on the tint. For cheapness, nothing is like smoked
glass; it is also said to intercept heat very perfectly, by Mr.
Prince, who places it within the eye-piece, close to the
"stop," or circular opening, which bounds the field; but thus
it can have only one degree of depth, and must be taken out
to view other objects. A strip of glass may be smoked to
different densities in different parts, and held between the eye
and eye-piece; but it should be protected from rubbing by a

* Dawes and Lassell think 2 inches aperture enough for perfect
safety.

† The value of complementary, or at any rate dissimilar, tints in pro-
tecting the eye was known before the telescope. Fabricius observed a
solar eclipse in 1590 "per duplex diversi coloris vitrum;" and Apian
speaks of them 50 years earlier.

similar strip of glass placed over it, and kept from touching by bits of card at the corners, the edges of the two strips being bound round with gummed slips of paper, or tape.* Where expense is not regarded, an optician will provide a delightful graduated screen with two wedges of glass, plain and coloured. In any case we must not begin with too faint a shade, but try the deepest first, and change it if necessary.

To bring the sun into the field, do not attempt to gaze at it through the finder; point the telescope till the finder shews it centrally on the hand, or a paper held behind it, or bring it to shine through the eye-piece *before* screwing on the dark cap. The thickness of any external screen will contract the field much, unless the eye is brought as close up as possible. In beginning, or ceasing, to look, move the eye sideways to or from its place, as a focus of heat is sometimes formed at a little distance outside the eye-piece, which ought never to be admitted into the eye.

With these precautions, there is no fear for an ordinary sight; but should the light or heat be still unpleasant, the aperture may be contracted as recommended for defective glasses; † or for a very sensitive eye, or a whole company at once, the image of the Sun may be received direct from the eye-piece, without any screen, on card. Choose a field large enough to take in the whole disc, and alter the focus till the image on the card is *quite* sharp, and at a convenient distance as to size; any spots then visible will be easily and fairly

* Gum-water, or mucilage, should be always made with *cold* water. It is far stronger, and keeps a long time without growing mouldy.

† Schwabe used a contracted aperture and light screen; Herschel I. preferred full apertures and deep screens, as Dawes does now, for sharper definition.

seen. And so will specks of dirt in the eye-piece; but these
may be detected by moving the tube, as the true spots will
keep their places in the image. If the eye-piece includes
only part of the Sun, do not mistake the edge of the field as
shewn on the card, for the Sun's limb ; both are circular, but
the latter only will move so long as the telescope is fixed. If
a circle of suitable size is drawn on the card, and crossed by
lines forming small squares, the image may be adjusted to
coincide with it, and the progress of the spots may be marked
and recorded day after day. Captain Noble has found that
plaster of Paris, smoothed while wet on plate glass, gives a
most beautiful picture; he fixes a disc of it inside the base of
a pasteboard cone, blackened inside, 1 foot long, and 6 inches
across the large end ; the small end being opened so as to fit
close on the eye-piece, with a hole in the side of the cone to
view the image.

All being arranged, we shall find four points especially
worthy of attention : 1, the dark spots; 2, the faculæ ; 3, the
mottled appearance ; 4, the transparent atmosphere.

1. The *Dark Spots*. These are not always visible ; the
disc is occasionally entirely free from them, but more frequently one or more will be in sight. Unless very small, they
usually consist of two perfectly distinct parts ; a dark
" nucleus," commonly very irregular in its outline, which
resembles, as Secchi remarks, the creeping of a very dense
luminous material over an extremely rough surface,—and a
surrounding " umbra," (or " penumbra,") a fainter shade with
an equally definite, but in general less angular, boundary.
The nucleus appears black from contrast, but is not quite so,
as is evident when Mercury passes across the Sun. Sometimes
it is slightly and unequally illuminated, as if overspread by a

thin haze, which Secchi compares to cirrus, or mare's-tail clouds, and finds to be the harbinger of its decrease and extinction : sometimes it is intersected by narrow veins of light. In Dawes's very ingenious solar eye-piece,* a sliding plate of metal contains a row of holes gradually decreasing in size, each of which may limit the field in turn, while the whole is insulated by ivory, so as to prevent the eye-piece from getting heated; thus the luminous part may be shut off, and the spot alone viewed with a very light screen-glass. In this way, he has detected in all large and many small nuclei, a perfectly black spot, or opening, of much smaller size. Secchi, who confirms this discovery, finds that holes in a glazed visiting-card answer well. The most diligent of solar observers, Schwabe of Dessau, has seen an occasional reddish-brown colour in the spots, in the immediate neighbourhood of others of the ordinary greyish-black, so as to preclude deception ; in one instance, three telescopes, and several by-standers, agreed as to this fact. Capocci, in 1826, perceived a violet haze issuing from each side of the bright central streak of a great double nucleus; and Secchi, during the eclipse, 1858, March 15, saw a rose-coloured promontory in a spot visible to the naked eye. Schmidt records many tints, chiefly violet nuclei and yellowish umbræ, especially as cast on paper ; but this excellent astronomer's observations seem hardly consistent enough on this one point to have much weight. The penumbra, which in most cases encompasses considerable nuclei, occasionally comprises a group of them, and frequently outlasts them, is made up, according to Schwabe, of a multitude of black dots usually radiating in straight lines from the nucleus.

* A similar idea had occurred to Professor Wilson, of Glasgow, in the last century.

Secchi, with greater optical power, finds these radiations to be alternate streaks of the bright light of the disc, and the dark ground of the nucleus. The penumbra, Herschel II. observes, occasionally shews "definite spaces of a second depth of shade;" usually it is darkest at the outside: sometimes it includes brilliant specks, or streaks, even close to the nucleus. Schmidt describes one of these insulated specks as the brightest portion at that time visible. These spots are of all shapes and sizes, up to enormous dimensions; the nucleus frequently surpassing the earth in magnitude. Schwabe, in 1843, saw one between 70 and 80,000 miles in breadth. The penumbra, especially of a group, is often much larger. Herschel II., at the Cape of Good Hope, estimated the area of one to be 3,780,000,000 square miles. Schwabe and Schmidt speak of groups which have extended across more than a quarter of the disc. One, observed by Hevel in 1643, is said to have occupied one-third of it. Spots exceeding 50″, Schwabe finds visible to the naked eye through a fog, or dark glass. Struve says that this is very unusual in recent times; but Schwabe has often recorded such instances, sometimes repeatedly in twelve months; and I have seen them on several occasions, without any especial attention. When thus perceptible they surpass the earth at least three times, — if conspicuous, much more. A gregarious tendency is obvious, and the groups are apt to be parallel with the solar equator. Herschel II. says that, if they converge, it will be towards the preceding side of the disc. They are absent from the poles, and infrequent at the equator, which is the hottest part of the globe; on each side of it are two fertile zones, reaching as far as 30° or 35° each way; sometimes they exceed these bounds. Schwabe saw one in 50°, La Hire in 70° of solar latitude.

The recent observations of Peters and Carrington tend to unsettle these limits, which may be subject to change. Schwabe finds that the W. members of a group disappear first, and fresh ones are apt to form on the other side, on which also are the greatest number of minute companions; that the small points are usually arranged in pairs; and that near the edge of the Sun, the penumbræ are much brighter on the side next the limb. Herschel II. saw the penumbræ often best defined on the preceding side, and Capocci (with a $7\frac{3}{4}$ inch Frauenhofer achromatic at Naples) found that the principal spot of a group leads the way, and that the nuclei are better defined in their increase than diminution. Peters and Carrington observe a remarkable tendency to divergence in adjacent nuclei. The extraordinary mutability of the spots will be easily recognised; frequently they are in continual change, varying from hour to hour, and even more rapidly. Biela has found spots disappear while he looked at them: Krone has observed them to form within a single minute: Schwabe saw a penumbra increase from $1'3''$ to $5'2''$ in 24^{h}. Capocci noticed the temporary reduction of a nucleus four times as large as the Earth, to the dimensions of Europe, "under his eyes:" an unfortunately vague expression, as the Académie des Sciences have remarked; but characteristic of that surprising fluctuation which must strike every observer.

The inquiry into their nature is very perplexing, from the absence of terrestrial analogies: the Sun evidently belonging to a wholly different and entirely unknown class of bodies. Professor Wilson of Glasgow's theory, modified by Herschel I., seems generally adopted, that the spots are openings in a brilliant envelope or "photosphere,"* through which we see, in the penumbra, a deeper and less luminous region, shining possibly (so Herschel I. thought), with the reflected light of the photosphere; and at a still greater depth, the dark body of the Sun, forming the nucleus. This view rests on the perspective appearance of the penumbra, when near the limb, which usually † is more contracted on the side next the eye; and the depression is corroborated by several observations of actual notches in the limb.‡ Herschel I. thought these openings might be caused by invisible elastic vapour, rising from the dark body of the sun, and expanding in its ascent: his son inclines to the idea that a transparent atmosphere above the luminous stratum may be subject in its equatorial regions, like that of the earth, to hurricanes, forcing their way downwards to the surface: and this view is strengthened by Dawes and Secchi: the former discovering a rotation in two nuclei, one revolving through 180° in 6ᵈ, the other through 70° in 24ʰ; the latter perceiving a spiral or whirlwind form in the penumbra and nucleus of a large spot. Something of the kind has again been very recently seen by Mr. Birt, with Slater's large achromatic, and it is a point which deserves careful attention. Secchi has revived Wilson's theory, that the penumbræ are the shelving sides of openings reaching down to the dark surface, and thinks them filled

* Schröter used this very appropriate and now universally-admitted term as far back as 1792.

† Schwabe and Schmidt have observed exceptions to this rule, but they are uncommon.

‡ A depression on a globe will disappear in profile, unless its breadth and depth are considerable; hence such observations are rare; but they are recorded by La Hire, 1703; Cassini, 1719; Herschel I., 1800; Dollond and others, 1846; Lowe, 1849.

with a dense atmosphere; but in that case they would grow darker inwards, contrary to observation. He calculates that their depth is about $\frac{1}{3}$ the semidiameter of the earth, or upwards of 1,300 miles. From the nature of the photosphere, we might conjecture that of the spots, were it not equally unknown. Schwabe, like Herschel I., infers an irregular distribution of luminous clouds: Capocci, a bright dry crust, evaporating into shining gas. Arago ingeniously rendered it probable that the light is not that of white-hot matter, solid or fluid, but resembles that of flame. Secchi and Henry have shown that the spots are relatively cool. Herschel II. deduces the partial removal of definite films, floating on a dark or transparent ocean, rather than the melting of mist or mutual dilution of gaseous media; the analogy of the Aurora Borealis has also been alluded to by him and his father, and the rapidly alternating opacity and transparency of some chemical mixtures near the point of saturation may be worthy of attention, in seeking for analogies. But we are far, as yet, from any adequate explanation, though these wonderful processes are increasing in interest, since there is now more than a suspicion that they influence the whole dependent system. The extraordinary perseverance of Schwabe * has shewn that

* The late lamented President of the Astronomical Society, Mr. Johnson, thus refers to the presentation of their Gold Medal to this observer:—" It was not...for any special difficulty attending the research, that your Council has thought fit to confer on M. Schwabe this highest tribute of the Society's applause. What they wish most emphatically to express is their admiration of the indomitable zeal and untiring energy which he has displayed in bringing that research to a successful issue. Twelve years, as I have said, he spent to satisfy himself—six more years to satisfy, and still thirteen more to convince, mankind. For thirty years never has the Sun exhibited his disk above the horizon of Dessau

the spots have regular maxima and minima, with a period of about 10, or, according to Schmidt, 11 years, which corresponds so exactly with the period of all magnetic variations, that both are now ascribed to the same unknown power, and the spots are no longer objects of mere curiosity, but indications of a mighty force, one of the prime laws of the universe.*

Concurrent authorities have justified our assuming that the spots are openings; yet observations exist, looking another way; and it may be well to insert them from their curiosity, as well as their being seldom referred to. Dr. Long, who published a Treatise on Astronomy in 1764, states that he, "many years since, while he was viewing the image of the Sun, cast through a telescope upon white paper, saw one roundish spot, by estimation not much less in diameter than our earth, break into two, which immediately receded from one another with a prodigious velocity." Dr. Wollaston says, "Once I saw, with a 12-inch reflector, a spot burst to pieces while I was looking at it. I could not expect such an event; and therefore cannot be certain of the exact particulars: but the appearance, as it struck me at the time, was like that of a piece of ice when dashed on a frozen pond, which breaks to pieces and slides on the surface in various directions. I was then a very young

without being confronted by Schwabe's imperturbable telescope, and that appears to have happened on an average about 300 days a-year. So, supposing that he observed but once a-day, he has made 9,000 observations, in the course of which he discovered about 4,700 groups. This is, I believe, an instance of devoted persistence (if the word were not equivocal, I should say, pertinacity) unsurpassed in the annals of astronomy. The energy of one man has revealed a phenomenon that had eluded even the suspicion of astronomers for 200 years!"

* Wolf hopes to demonstrate that they result from a reaction of the planets upon the Sun.

astronomer, but I think I may be sure of the fact." Something not absolutely unlike has been recorded since. 1849, Feb. 26, two persons at Lawson's observatory repeatedly noticed, at regular intervals of 15^s, the alternate separation and reunion of a large nucleus; the next day it had become permanently divided, but the smaller portion was undergoing the same process at intervals of 1^m, and when these ceased the interstice between the two spots varied in breadth in a similar way: another spot subsequently shewed the same oscillations. Robinson, also, at Armagh, witnessed a bridge of penumbral matter shot across a nucleus through some thousands of miles in a few minutes. From such appearances, an observer, unacquainted with the ordinary theory, might possibly have inferred the solidity, from the disruption, of the dark object.

Notwithstanding their changeable nature, the larger spots are possessed of some permanency. After describing straight lines about June 11 and Dec. 12,* but elliptical paths at other times, in consequence of the position of the Sun's equator towards our eye, they go out of sight at the W. limb, and if not dissipated, return at the E. edge after about 12$\frac{1}{2}$^{d} to run the same course. Some have thus continued through many revolutions. In 1779, a large spot continued visible for 6 months, and in 1840 and 1841, Schwabe observed 18 returns (though not consecutive), of the same group: the most permanent, he says, are usually round, of a moderate size, and not sharply defined. There has been a suspicion that after long intervals they may re-appear in the same places; but it seems unlikely in the face of the now-

* So Herschel II., Outlines of Astronomy, ¶ 390. But in the next paragraph they stand as July 12 and Dec. 11.

established fact of their proper motion. Fritsch has stated
that he has seen one stand nearly still for three days, and
Lowe, that he has even witnessed retrogradation;—assertions
which may involve a suspicion of mistake. Schröter and
others have ascribed to them a more moderate locomotion.
This was micrometrically established in a lateral direction by
Challis, in 1857; and Carrington has recently made known
his very interesting discovery, that there appear to be currents
in the photosphere, drifting the equatorial spots forward, in
comparison with those having considerable latitude; with
lateral deviations of smaller amount. With these shifting
landmarks, it is not surprising that the Sun's period of rotation is still doubtful; though Laugier's value, $25^d 8^h 10^m$,
cannot be far wrong. Relative displacement in groups would
be an interesting study, requiring neither micrometer nor clock,
—only careful drawing. Several observers have found that
spots near the limb require a different focus from those in the
centre: this is no doubt, as Dawes says, an optical deception.

2. The *Faculæ*, or bright streaks. Less obvious than the
dark spots, and requiring more power, these are not difficult
objects; to be looked for in the usually spotted regions, but
only near the limbs. They are irregular, curved, and branching, considerably more luminous than their vicinity, but not,
according to Secchi, than the centre of the Sun. They are
proved to be what they appear, ridges in the photosphere, by
an observation of Dawes, who saw a facula project above the
limb as it turned across it into the other hemisphere : but we
seldom find them visible close to the limb, or far from it, as
they are changed in the centre into bright tufts and specks,—
an effect of perspective, as Secchi has pointed out, their
breadth being much less than their height. They are as vari-

able as the spots, and probably depend upon them, surrounding them usually near the limb, and sometimes as they cross the centre, attending their development, and succeeding their dissolution. Secchi compares them to immense waves raised by the outburst of the spots, of which they are commonly the harbingers. They have been supposed to rise so high as to form the red prominences, the wonders of total solar eclipses; but this is very doubtful.

3. The *Mottled Surface*. An attentive eye will soon detect a marbled or curdled appearance over the whole Sun, even to the poles, formed by a mixture of small patches of white and grey. Herschel I. refers it to a double stratum of clouds, the lower (grey) closely connected; the upper (white) chiefly detached; the white forming faculæ by aggregation; the grey passing into the condition of penumbræ, and often containing black points, the occasional germs of nuclei; but, as he remarks, the grey part cannot be much depressed, since it remains visible, notwithstanding much perspective foreshortening. The earliest mention I have noticed of this mottling is during the solar eclipse, 1748, July 14, (o. s.) when it was clearly described by Mr. J. Short, (the eminent optician?) and was quite new to him: since that time it has been remarked by all careful observers. Herschel I. missed it once only; the Sun being quite uniform, 1795, July 5. Herschel II. cannot see it with any *achromatic*, but this must be a peculiarity, as in mine it was usually an easy object; it is occasionally strongest in the zones marked by the spots, where Schwabe saw it in 1831, like two freckled girdles round the Sun. He has sometimes observed the grey parts variegated by minute dark veins. Secchi, with Dawes's eye-piece, finds the surface as rough as a tempestuous sea, or a newly-ploughed field. Should this freckled

appearance not be at once detected in a clear sky, a slight shaking of the image by tapping the telescope may render it perceptible.

4. The *Transparent Atmosphere.* That this exists, and is of far greater proportionate extent than planetary atmospheres, is made apparent in total eclipses by the luminous corona and red prominences which then surprise the spectators; but these occasions in England are so distant in prospect that a passing mention of them may suffice. The presence and great density of an atmosphere is however evident from the faintness of the limb compared with the centre of the disc. This was early detected at Rome by Luca Valerio, called by Galileo the Archimedes of his time; it was denied by the inventor of the telescope, but may be easily perceived when the image is cast on paper.* By means of a doubly-refracting prism, Secchi has found the limb not more luminous than the penumbræ of spots in the middle of the Sun, that is, deprived of about half the light; and his previous discovery of a similar decrease of heat confirms the existence of such an absorbent envelope. Hence also may result the inferior distinctness of the solar limb as compared with that of the Moon in eclipses, a fact remarked by Airy, and exhibited by photography; the focus of the objects is however not the same.

Something must be said of the Sun as a background for the exhibition, in a fresh aspect, of intervening bodies. Partial solar eclipses seldom gain much interest in the telescope, excepting from the occasional projection of mountains on the Moon's limb. The total eclipse is so wonderful and so fugitive as to

---

* Arago's polariscope fails to exhibit this obvious difference; and hence some doubt is thrown upon its indication of the gaseous nature of the photosphere.

require special preparation, and is too rare in England to require
notice here. The transits of Mercury will be found under the
head of that planet : those of Venus are too uncommon and
distant to be admitted in our limited space; the next visible
in England not occurring till 1882, its successor not till 2004.
Comets, from the endless diversity of their paths, must occasionally pass over the Sun ; and the opportunity of such a
background would be most valuable for acquiring more information as to their nature. This actually took place, 1819,
June 26; but it was not known till afterwards, when a
controversy arose as to whether the stranger had been visible.
Pastorff maintained that he had that day seen on the Sun a
round dark nebulous spot with a bright point in the centre :
Olbers and Schumacher thought there was no ground for supposing it to have been the comet. In 1826, Gambart found
that a comet which he had discovered would be on the solar
disc a little after its rising on Nov. 18, and gave immediate warning to the astronomers of Europe ; but the weather
caused almost universal disappointment; only Gambart and
Flaugergues had a view of the Sun, and neither of them could
perceive a trace of the comet. Yet it seems not impossible
that nuclei as vivid and brilliant as those of 1744 and 1843
might stand out in some kind of relief; and therefore if anything unusual should ever be noticed on the disc, it should be
carefully watched; and should its rate of progress shew that
it is not an ordinary spot, its appearance ought to be most
critically examined with various powers and screen-glasses,
and intelligence sent instantly to the principal observatories
within reach, that the comet, if such, may be re-discovered as
soon as possible.

But there is another reason for attention to anything of un-

wonted aspect on the Sun : it appears that opaque bodies, *not*
comets, and of most mysterious character, traverse it from time
to time. 1798, Jan. 18, the Chevalier D'Angos, at Tarbes,
perceived a black, well-defined, somewhat elliptical spot on the
Sun, about half way from centre to limb, which he was soon
astonished to find in motion, and which he watched carefully
till it left the disc 20^m afterwards. He had been surprised in
1784 at the disappearance of a round spot which was on the
Sun some hours before, but he did not see it actually *in motion.*
The Chevalier's reputation for accuracy does not indeed stand
high. Smyth tells us, that in commemoration of some strange
mistakes, Baron de Zach "was induced to term any egregious
astronomical blunders—*Angosiades.*" Still there is something
so circumstantial in the story of 1798, that it is difficult to
withhold our belief; especially as corroborative evidence may
be found. Fritsch, at Quedlinburg, who thought he perceived
many proper motions in the solar spots, saw one, 1800, March
29, small, black, and almost without penumbra, which increased
its distance from the limb 18$\frac{1}{4}$' in 6^h : another spot, 1802,
Feb. 27, showed rapid and anomalous motion ; another,
1802, Oct. 10, had moved 2' in 3^m, and, after a cloudy
interval of 4^h, had disappeared. The next instance is so well
attested that it seems a pity to curtail it; it is thus given by
Mr. Capel Lofft, of Ipswich, in a letter to the Monthly Maga-
zine, dated 1818, Jan. 10. "I saw it about 11, A.M." (on
Jan. 6,) "with my own reflector, with a power of about
80 ; with an excellent Cassegrain reflector made by Crickmore
of this town, with about 260 ; and with a reflector of Mr.
Acton's, with about 170. It appeared, when I first saw it,
somewhat about one-third from the eastern limb, subelliptic,
small, uniformly opaque. About 2$\frac{1}{2}$ hours, P. M. it appeared

to Mr. Acton considerably advanced, and a little west of the
Sun's centre; and I think it appeared then 6 or 8 seconds in
diameter. I had been able to see no spot on the 4th, nor again
on the 8th, and even on the 6th Mr. Crickmore could not see
it a little before sunset, though the telescope already mentioned
gave him every advantage. Its apparent path while visible
seemed to make a small angle with the Sun's equator. Its
state of motion seemed inconsistent with that of the solar
rotation, and both in figure, density, and regularity of path,
it seemed utterly unlike floating scoria. In short its progress
over the Sun's disc seems to have exceeded that of Venus in
transit. There are two instances, if not three, of Comets seen
in transit, and this phænomenon seems to have been one. I
wish it may have been seen elsewhere." Of the comets here
mentioned I have never found any record; and it is more
probable that this strange object possessed greater density than
any of those nebulous bodies. A less satisfactory instance is
cited in 1833, when Pastorff saw a small round spot so frequently on the Sun, that he suspected its planetary nature; but
it does not appear that he marked its rate of progress, and it
was unnoticed by other observers.— 1845, May 11, 12, 13, a
singular spectacle presented itself to Capocci and other observers
at Naples, with various telescopes; one, a large achromatic by
Cauchoix. A number of well-defined dark bodies were seen
from time to time crossing the Sun's disc, from $1''$ to $2'$ or $3'$
in size, all circular, but the larger ones with irregular protu-
berances, in paths usually but not invariably straight, generally
parallel, but occasionally deviating in all possible angles, with
very varying rapidity, and, as appeared by the altered focus, at
very different distances from the eye. During 10^m, 102 were
counted, but not more than four or five at once. In this strange

D

phænomenon Erman saw a strong confirmation of his previous conjecture, that a mass of the meteors called Shooting Stars would pass between the Sun and Earth about this time ; and the coincidence is at any rate very curious. Two such spots were also seen at Deal at another of the meteoric periods, February, 1849. On two occasions, in 1847 and 1849, Schmidt was witness to the rapid passage of a small black body, neither insect nor bird, across the solar disc.* The reader will excuse the room allotted to these somewhat unfamiliar and certainly very singular details.

MERCURY.

THIS planet, though at times readily visible to the naked eye, is but seldom seen from its nearness to the Sun ; and often lies too near the horizon for the telescope. A well-adjusted equatorial stand will find it by day, but its small diameter of less than 3,000 miles subtends at a mean not more than 6″ or 7″ ; and ordinary observers will not see much, where professed astronomers have usually found little. But as these pages may possibly fall into the hands of some, whose advantages or enterprise may lead them to attack a neglected object, the following points may be specified.

1. The *Phases*. These will be easily seen, and are only remarkable because the breadth of the enlightened part has been sometimes found less than it should be from calculation.

---

* An examination of the original memoir shews that an observation by Messier, 1777, June 17, which has been thought of a similar nature, points rather to the intervention of drops of rain or hail-stones.

Schröter noticed this; and it is confirmed by Beer and Mädler: but their explanation of a dense atmosphere making the terminator, or boundary of light and darkness, faint, is inadequate, as their observation was before sunrise, when the dullest part of the disc would still be very luminous. See this again in Venus.

2. The *Mountains*. At the close of the last and beginning of this century, Schröter, of Lilienthal in Hanover, a most diligent observer, and his assistant Harding, obtained what they deemed sufficient evidence of a mountainous surface in the occasional blunting of the S. horn, some minute projections on its outer edge, and an irregular curve of the terminator; they gave the inferred elevations a height of nearly 11 miles perpendicular.

3. The *Atmosphere*. The decrease of light towards the terminator, and the occasional presence of dark streaks and spots, indicated to the same astronomers a vaporous envelope, where they thought they even saw traces of the action of winds. From a combination of these appearances they deduced a rotation in 24^h 0^m 53^s on an axis inclined about $70°$ to its ecliptic. But further observations are needed. In De La Rue's magnificent Newtonian, 10 feet focus and 13 inches aperture, constructed by himself, the planet has a rosy tinge.

Transits of Mercury are comparatively frequent; they will be visible in Europe in 1861, 1868, 1878, 1894. The planet breaks in upon the Sun as a dark notch, sometimes preceded, it is said, by a penumbral shade; but the earliest impression will be missed, unless the exact point of the Sun's limb is known, and kept central in the field. As it advances, the part of Mercury not yet entered on the Sun may be rendered visible by being projected upon the "corona" or illuminated atmosphere, which is so conspicuous in total solar eclipses, and has been known to relieve dark bodies in front of it, such as Mercury, Venus, or even a portion of the Moon. On finally entering the Sun, or beginning to leave it, the planet has been seen lengthened towards the limb; probably from irradiation, which often enlarges luminous images at the expense of contiguous dark spaces. Fully on the Sun, Mercury appears intensely black; some astronomers have given it a slight dusky border, others a narrow luminous ring; both, probably, deceptions from the violent contrast and the fatigue of the eye, especially as others have seen neither. But whatever is seen should always be recorded. A stranger appearance is better attested—that of a whitish or grey spot on the dark planet, seen by Wurzelbauer, 1697, Schröter, Harding, and Kohler, 1799; Fritsch and others, 1802; Moll and his assistants, 1832 (when Harding clearly distinguished two spots, and Gruithuisen suspected one); and recognised in England and America, 1848. No terrestrial analogy will explain a luminosity thus visible close to the splendour of the Sun; but the testimony seems irresistible. Schröter and Harding ascribed to these spots a motion corresponding with the rotation which they subsequently inferred from other indications. *

* A similar phænomenon was observed on Venus in the transit of 1761 (Append: ad Ephem: Astron: 1766, 62), for the explanation of which, in the additional light derived from the transits of Mercury, the Abbat Hell's theory of optical illusion seems quite insufficient.

VENUS.

Fairest of stars, last in the train of night,
If better thou belong not to the dawn,
Sure pledge of day, that crown'st the smiling morn
With thy bright circlet, praise HIM in thy sphere.

MILTON.

THE most beautiful of all the heavenly bodies to the unaided eye is often a source of disappointment in the telescope. For the most part it resists all questioning beyond that of Galileo, to whom its phases revealed the confirmation of the Copernican theory ; an important discovery, which he involved for a season in the following ingenious Latin transposition :

Hæc immatura à me jam frustra leguntur, o. y.

the letters in their original order forming the words

Cynthiæ figuras æmulatur mater amorum.

Observers in general will subscribe to the experience of Herschel II., who says it is the most difficult of all the planets to define with telescopes. " The intense lustre of its illuminated part dazzles the sight, and exaggerates every imperfection of the telescope ; yet we see clearly that its surface is not mottled over with permanent spots like the Moon ; we notice in it neither mountains nor shadows, but a uniform brightness, in which sometimes we may indeed fancy, or perhaps more than fancy, brighter or obscurer portions, but can seldom or never rest fully satisfied of the fact ; " and he infers, like his father, and Huygens long before, that " we do not see, as in the Moon, the real surface of these planets " (Venus and Mercury) " but only their atmospheres, much loaded with clouds,

and which may serve to mitigate the otherwise intense glare
of their sunshine." The conclusion, however, is premature;
perseverance has triumphed, though only in the pure Italian
sky; and moderate-sized telescopes have in some cases detected
what magnificent instruments had failed to reveal. The
following are the chief points of interest :—

1. The *Phases.* These are easily seen, and very beautiful,
excepting the dull gibbous form in the *superior* or further
part of the orbit, where the disc is also small; near the
greatest elongation from the Sun towards E. when it is an
evening, or W. when a morning star, Venus puts on a beautiful
shape—that of the moon in quadrature; between these points
in the *inferior* or nearer part of the orbit, she is a lovely
crescent, larger, sharper, and thinner in proportion as she is
nearer really to the Earth and apparently to the Sun. This
crescent has been seen even with the naked eye in the sky of
Chile, and with a dark glass in Persia, and a very small
telescope will shew it. When quite slender it should not be
looked for after sunset or before sunrise, as it lies too low in
the vapour; but an equatorial stand will find it in the middle
of the day—a lovely object; and thus, or in a transit instrument, it has been traced as a mere curved thread extremely
near the Sun.* Ordinary observers may succeed in seeing it
a very delicate crescent soon after it has passed its inferior
conjunction, by watching for its earliest appearance

"Under the opening eyelids of the morn,"

setting the finder, or the telescope with the lowest power,
upon it, and following it at intervals sufficient to keep it in

* Vidal is said to have seen Venus only 2′ from the Sun's limb; but
this must have been in its upper, or *full moon* conjunction.

the field, till it has cleared the vapours of the horizon ; in this way it may be readily viewed for hours. In fact, the planet is best seen for many purposes in the day-time ; its light, unpleasantly dazzling in a dark sky, so as to bear a screen-glass, is subdued by day to a beautiful pearly lustre. Nor is it very difficult to find. For some time about its greatest brightness, at 40° from the Sun in the inferior part of the orbit, it not merely casts a shade by night, but is visible to the naked eye at noon-day, provided its position is pretty well known. A little careful steady gazing will then bring it out as an intense white point, and, if the air is good, it will be a charming telescopic object. At other seasons, a little hand telescope with a large field will shew it much sooner in the evening or later in the morning than might have been expected. Like that of Mercury, the phasis often disagrees with calculation ; at its greatest elongations it ought to be exactly " dichotomized" as a half-moon ; but in August, 1793, Schröter found the terminator slightly concave in that position, and not straight till eight days afterwards ; and Beer and Mädler fully confirmed this by many observations in 1836, proving that the apparent " half-moon" takes place six days earlier or later than the computed, according to the direction of the planet's motion. They found also that there is a similar defect in the breadth of the crescent. Neither their explanation of this, through the shadows of mountains, nor Schröter's, through the fading of light towards the terminator, is satisfactory as far as night observations are concerned. In 1839, De-Vico and his assistants at Rome found a similar discrepancy of about three days. This curious phænomenon is so easily seen, that I perceived it with an inferior fluid achromatic, on Barlow's plan, 1833, March 6, before I knew that it had been noticed by others.

2. The *Mountainous Surface.* La Hire, in 1700, with an old 16 feet refractor, power 90, professed to see irregularities in the crescent when very narrow; and Briga not only described an indented terminator, but unjustifiably altered Cassini's figures accordingly. But Schröter's observations at the end of the last century are far more trustworthy. With several fine reflectors and a very good achromatic, he and others found the boundary sometimes slightly jagged, sometimes irregular in curvature, so as to vary the relative thickness of the horns; and these would occasionally pass through such changes as to show a rotation in 23^h 21^m 8^s. At the quadrature, the N. cusp would frequently project, while the S. was blunted, with sometimes a minute point cut off from it by a narrow black line, the shadow apparently of a lofty ridge; some of these mountains Schröter supposed to be 27 or 28 miles high, but of course with great uncertainty. Herschel I. attacked these discoveries in the Philosophical Transactions for 1793, in what Arago justly terms " une critique fort vive, et, en apparence du moins, quelque peu passionnée." Schröter calmly and satisfactorily vindicated himself through the same medium in 1795; and Beer and Mädler, in 1833 and 1836, have fully established his accuracy as to an irregularly curved terminator, causing great and rapid changes in the shape of the cusps. They also saw an occasional bending off of the S. horn, corresponding with Schröter's flattening of the limb in that direction.* Fritsch, in 1799 and 1801, witnessed several of

---

* It is a pity that the figures in their "Beiträge" are so bad. Schröter's, though not good, are more intelligible. The Roman observers have also noticed this strange temporary flattening of the circular limb near the S. horn, and consider it a profile view of one of the large grey spots, which, if so, must be deeply depressed.

these appearances with a little achromatic of $2\frac{1}{2}$ feet focus. Flaugergues and Valz noticed an irregular terminator, but deny its changes. Breen, with the great Northumberland telescope at Cambridge, of $19\frac{1}{2}$ feet focus, and $11\frac{1}{2}$ inches aperture, has often seen the unequal curve of the terminator, and the blunted S. horn. But the most curious observations are those made by De-Vico and his assistants at Rome, in April and May, 1841. An achromatic by Cauchoix, $6\frac{1}{4}$ inches aperture, with powers sometimes up to 1128, enabled them to trace the approach towards the terminator, night after night, of a valley surrounded by mountains like a lunar crater, $4\cdot5''$ in diameter. The crescent was narrow, and near the N. horn they first saw an oblong black spot, which subsequently was bordered with stronger light, afterwards encroached with half its ring upon the dark side, and ultimately formed a black notch between two bright projections, giving the appearance of a triply-pointed horn; a longer black streak was seen near the other cusp at the same time.

3. The *Spots*. These have occasioned much controversy, from their indistinctness, which is such that the Virgilian "aut videt, aut vidisse putat" is often the observer's conclusion. There have been, however, many exceptions. In 1666 and 1667, Domenico Cassini, at Bologna, repeatedly saw one bright and several dusky spots, the former giving a rotation in 23^h 21^m; subsequently, in the air of France, he never could rediscover them. In 1726 and 1727, Bianchini, at Rome, observed spots repeatedly with a 66 feet refractor, aperture a little more than $2\frac{1}{2}$ inches, power 112, like the "seas" in the moon to the naked eye, though less distinct, and not till after sunset, from want of light in his glass. His figures shew them surrounding the equator of Venus, forming three oceans as it

were, one tolerably circular, the others much lengthened, and
each of the latter subdivided by narrow straits into three portions; besides two spots, one occupying the S. polar region, the
other a horse-shoe round the N. pole. He did his work well,
but for want of *sky-room*, did not perceive their rapid motion,
and gave a wrong rotation of 24ᵈ 8ʰ. Schröter and Herschel I.,
half a century later, with much finer instruments, but in less
pellucid skies, could only make out through many years a few
faint markings, with suspicions of motion; at last, 1801, Aug.
29, Schröter detected a dim oblique dusky streak, like one on
Mercury, giving a rotation in about 24ʰ. Several other astronomers have had occasional glimpses of darker shadowings;
Gruithuisen speaks of minute brilliant round specks, and
Schumacher perceived a dusky spot in the twilight, which in
half an hour was lost in increasing glare; and found others,
visible with a small telescope, effaced in a larger one—a caution
for future observers. But the most effective results are those
of De-Vico at Rome, 1839-1841, who had been instigated by
Schumacher to verify Bianchini's assertions in the same atmosphere. He used the Cauchoix achromatic chiefly by day, since
in a dark sky the glare overpowered the spots so much as to
render micrometrical measures useless, and account for Bianchini's erroneous rotation; the drawings of the latter were
however found, with the omission of one small spot, remarkably exact. Of six observers, the most successful in seeing
these faint clouds were those who had most difficulty in catch-
ing very minute companions of large stars; * but all agreed in

---

* De-Vico assigns no reason, but it is obvious enough, and worthy of
notice. A very sensitive eye, which would detect the spots more readily,
would be more easily overpowered by the light of a brilliant star, so as
to miss a very minute one in its neighbourhood. There is abundant

the figures, and witnessed a progression, giving a rotation in
23^h 21^m 22^s on an axis greatly inclined, though less so than
Bianchini had supposed. Could we but see these spots more
readily, what an interesting object would this lovely planet
become, especially as in point of size it is the only companion
to the Earth in the whole system! And the possessors of even
common telescopes need not despair, though their chances may
not be great. Were the difficulty in the atmosphere of the
planet, as has been supposed, the case would be more unpromis-
ing; but a curious instance, concurrent with the results of the
Italian sky, will prove that the impediment lies nearer home,
and that some favourable conjuncture may yet be in store for
us, especially remembering that at Rome the spots have been
seen even with a little telescope of 2 inches aperture. The
story is so curious that it must be given entire from the
Philosophical Transactions : " January 23rd, 1749-50, there
was a splendid Aurora Borealis. About 6^h P.M., the Rev.
Dr. Miles, at Tooting, had been viewing Jupiter and Venus,
and shewing them to some friends, with one of Short's
reflectors, greatest power 200, when a small red cloud of the
Aurora appeared, rising up from the SW. (as one of a deeper
red had done before), which proceeded in a line with the
planets, and soon surrounded both. Venus appearing still in
full lustre, he viewed her again with the telescope, without
altering the focus, and saw her much more distinctly than

industry in these Roman observations : Palomba, the assistant, made
11,800 measures in 1839, of which 10,000 were employed in determining
the rotation. But De-Vico's style wants explicitness, and there are
strange traces of inexperience or inattention in the Jesuit College, rendering the memoirs of that date not quite satisfactory. The present
Director, Secchi, is a man of a very different mould.

ever he had done upon any occasion. All his friends were of
the same opinion as to the sight they had of her on that occasion.
They all saw her spots plain, resembling those in the Moon,
which he had never seen before, and this while the cloud
seemed to surround it as much as ever; but whether the
vapour might be rarer nearer the planet, no judgment could
be made, because of her too powerful light."

4. The *Atmosphere*. The bright border noticed by some
observers as attending the circular limb may be a deception;
but there is very sufficient proof of the existence of a vaporous
envelope. Schröter's dusky belt, already mentioned, indicates
it.   He has ascribed to it the great decrease of light towards
the terminator and cusps; and he and Herschel I. agreed as to
the extension of the horns beyond a semicircle, which may be
due in part to the penumbra, or additional daylight caused by
the Sun's not being a point but a great disc, but more to refraction through an atmosphere.   Schröter also perceived a faint
gleam along the limb beyond the horns, a true twilight, produced by an atmosphere which must be somewhat denser than
our own.   In May, 1849, Mädler found the horns projecting
to 200° and even 240°, shewing a refraction about $\frac{1}{6}$ stronger
than ours; and hence Cassini in 1692, and Drew in 1854,
found the crescent too broad near the conjunction.   Secchi, in
1857, saw in that position the cusps much prolonged, and the
twilight extending $19\frac{1}{2}°$, even through our strongly illuminated
atmosphere.

5. The *Phosphorescence of the Dark Side*. This truly
unaccountable appearance* is remarkably well attested. It is

* Arago's " negative visibility" is a very perplexing attempt at solution. The faint illumination which renders some of our terrestrial

noticed as far back as 1715, in the "Astro-Theology" of Derham, who says that "this sphæricity, or rotundity is manifest in our Moon, yea and in Venus too, in whose greatest Falcations the dark part of their Globes may be perceived, exhibiting themselves under the appearance of a dull, and rusty colour." 1720, June 7, Kirch, junior, believed that he saw it, the crescent being then extremely narrow. 1759, Oct. 20, it was seen by Andreas Mayer, through merely a transit instrument, and, strange to say, at 44ᵐ after noon : "etsi pars lucida Veneris tenuis admodum erat, nihilominus integer discus apparuit, instar lunæ crescentis, quæ acceptum à terrâ lumen reflectit." Schröter and Herschel I. perceived traces of it. Von Hahn says he saw it repeatedly, by day as well as night, and with several instruments; he was, however, an inferior observer. In 1806 it displayed itself beautifully to Harding three times, and to Schröter once, within five weeks. Pastorff also witnessed two of these especial appearances. Guthrie and others noticed it a few years ago, with small reflectors, in Scotland ; Mr. Purchas, at Ross, in England ; De-Vico and Palomba, often, in Italy. I have suspected it more than once. The dark side is frequently too small in proportion, like that of the crescent Moon to the naked eye ; and from the same cause,— the irradiation of the luminous part; it is sometimes described as grey, sometimes reddish. It would be well worth looking for when the crescent is narrow, but Venus should have high N. latitude to clear the vapours of the horizon.

6. The *Satellite*. This is an astronomical enigma. It is not easy to set aside the evidence of its occasional appearance,

nights lighter than others, remarked by Schröter, Arago, and, I think, by myself, scarcely affords an analogy. Humboldt gives a striking instance, Cosmos I., 131 (Bohn).

yet, if it exists, it should be always visible, for the diameter
ascribed to it is about $\frac{1}{4}$ that of Venus. Cassini saw it in 1672
and 1686; Short in 1740; * Mayer in 1759; Montaigne in
1761 ; Rödkier, Horrebow, and three others, at Copenhagen,
and Montbarron, at Auxerre, in 1764. The Abbat Hell
maintains that all these were images formed by reflection in
the eye-piece, in which way he could produce a satellite at
will. No doubt some of the observations may be thus disposed of, but not, at any rate, Short's testimony, who, by
using two telescopes, and at least four eye-pieces, rendered the
Abbat's conditions for the illusion almost impossible. Humboldt classes it with the " myths of an uncritical age."
Smyth inclines to an opposite opinion. Hind considers it "a
question of great interest," and says it "must remain open
for future decision."

Mädler tells a strange story about a number of brushes of
light diverging from the circular side of the crescent Venus,
lasting as long as the planet could be seen that evening, and
remaining unaffected by any turning round or change of the
eye-piece. He attempts no explanation, but thinks it could
not have been an optical illusion. This is certainly *possible*,
but it is an instructive instance of the oversights which may
be incidental even to great philosophers, that it never seems to
have occurred to him to try another telescope !

———◆———

THE MOON.

THE comparatively small distance of our satellite, 240,000
miles, renders it the easiest of telescopic objects. Its shadowed

* The air was then so clear that two darkish spots were visible.

and irregular surface, visible to the naked eye, is well brought
out even with a low magnifier; hence Galileo readily com-
prehended the nature of what his new and imperfect invention
disclosed to him, and the smallest instruments will shew that
freckled aspect, arising from numberless craters, which he compared to the eyes in a peacock's tail. Many a pleasant hour
awaits the student in these wonderful regions; only, let him
not expect that what he sees so plainly will be equally
intelligible, excepting in its unquestionable relief from the
effect of light and shade. Very overstrained ideas, as to the
possibility of making out the minute details of the surface,
have been entertained, not much more reasonable than those
of the islanders of Teneriffe, whose simplicity led them to
imagine that the telescope of Piazzi Smyth would shew their
favourite goats in our satellite; a very little consideration,
however, will detect the absurdity of such anticipations.
The "Moon Committee" of the British Association have
recommended a power of 1,000; few indeed are the instruments or the nights that will bear it; but when employed, what
will be the result? Since increase of magnifying is equivalent
to decrease of distance, we shall see the Moon as large (though
not as distinct) as if it were 240 miles off, and any one can
judge what could be made of the grandest building upon
earth at that distance: very small objects, it is true, are
perceptible from their shadows, but their nature remains
unknown. Much difficulty too arises from the want of terrestrial analogies. It may be reasonably supposed that Venus or
Mars, at the like distance, would be far more intelligible.
We should probably not find them perfect transcripts of our
own planet, for, as Schröter often remarks, variety of detail in
unity of design is characteristic of creation; but we should

have a fair chance of understanding much of what we saw. It is quite otherwise with the Moon. It is, in Beer and Mädler's words, no copy of the Earth; the absence of seas, rivers, atmosphere, vapours, and seasons, bespeaks the absence of "the busy haunts of men;" indeed of all terrestrial vitality, unless it be that of an insect or reptile. Whatever may be the features of the averted hemisphere, which, as Gruithuisen and Hansen have suggested, may exist under other relations, on this side we perceive but an alternation of level deserts and craggy wildernesses, all barren, and cold, and dead. The hope which cheered on Gruithuisen and others, of discovering the footsteps of human intelligence, must be abandoned. If it should be hereafter found—and it is not impossible—that the lunar surface is habitable in some way of its own, we have reason to suppose that, where the conditions of life are so extremely dissimilar, its traces would be as undecipherable by our experience as a brief inscription in a character utterly unknown. We ought not, in fact, to be surprised at such a difference between bodies belonging to distinct classes. It would have been unreasonable to have looked for a duplicate of a primary planet in its attendant; and, waiving any disappointment from this cause, we shall find the Moon a wonderful object of study. It presents to us a surface convulsed, upturned, and desolated by forces of the highest activity, the results of whose earliest outbreaks remain, not like those of the Earth, levelled by the fury of tempests, and smoothed by the flow of waters, but undegraded from their primitive sharpness even to the present hour. The ruggedness of the details, we are assured, becomes more evident with each increase of optical power, and we cannot doubt that we look upon the unchanged results of those gigantic operations which have stereotyped their record on nearly every region of the lunar globe.

A brief general description of the phænomena of the Moon will prepare us for an examination of its topography. We have then

1. The *Grey Plains*, or *Seas* as they were formerly believed to be, and are still termed for convenience.* These are evidently dry flats—if the term "flat" can be applied to surfaces shewing visibly the convexity of the globe—analogous to the deserts and prairies and pampas of the Earth. Beer and Mädler find that they do not form portions of the same sphere, some lying deeper than others : they are usually of a darker hue than the elevated regions which bound them, but, with a strong general resemblance, each has frequently some peculiar characteristic of its own.

2. The *Mountain Chains, Hills*, and *Ridges.* These are of very various kinds : some are of vast continuous height and extent, some flattened into plateaus intersected by ravines, some rough with crowds of hillocks, some sharpened into detached and precipitous peaks. The common feature of the mountain-chains on the Earth, a greater steepness along one side, is very perceptible here, as though the strata had been tilted in a similar manner. Detached masses and solitary pyramids are scattered here and there upon the plains, frequently of a height and abruptness paralleled only in the most craggy regions of the Earth. Every gradation of cliff and ridge and hillock succeeds :

* Riccioli, when he recast the lunar nomenclature, and substituted the names of philosophers for the feeble geographical analogies of Hevel, retained the generic title of " seas," though he altered their designations. The reform attempted by Gruithuisen, who would have had them called " surfaces," has never taken effect.

E

among them a large number of narrow banks of slight
elevation but surprising length,* extending for vast distances
through level surfaces : these so frequently form lines of communication between more important objects,—uniting distant
craters or mountains, and crowned at intervals by insulated hills,
— that Schröter formerly, and Beer and Mädler in modern
times have ascribed them to the horizontal working of an elastic
force, which, when it reached a weaker portion of the surface,
issued forth in a vertical upheaval or explosion. The fact of
the communication is more obvious than the probability of the
explanation.

3. The *Crater-Mountains*, comprising both the ridge and
the included cavity. These are the grand peculiarities of the
Moon : commonly, and probably with correctness, ascribed to
volcanic agency : yet differing in several respects from the
foci of eruption on our own globe : on the Earth, they are
usually openings on the summits or sides of mountains—on the
Moon, depressions below the adjacent surface, even when it is
a plain or valley ; on the Earth, the mass of the cone usually
far exceeds the capacity of the crater—on the Moon, they are
much nearer equality ; on the Earth, they are commonly the
sources of long lava-streams—on the Moon, traces of such out-
pourings are rare ; on the Earth, their dimensions are comparatively inconsiderable—on the Moon, many of them are, the
grey plains excepted, among the largest of its features. When,
however, allowance has been made for the inferior power of
gravity on the Moon, through which a six-fold displacement in
height or distance would be caused by the same amount of
force, — for the probable difference of materials, — and for the

---

* Schröter gives a length of 630 or 640 miles to a ridge connecting
the spots Copernicus and Kirch.

more rapid cooling produced by radiation in the absence of an atmosphere, it is certainly possible that volcanic force, similar to that on the Earth, may have been the real agent, though in a greatly modified form. Any one may see, with the ingenious Hooke, a strong resemblance to the rings left by gaseous bubbles; but to this impression mechanical difficulties arising from the cohesion of materials have been opposed, and a more consistent explanation sought in the idea that the larger craters may be the remains of molten lakes; in these, left for a while unfrozen in the general cooling and crusting over of the once-fiery globe, an alternate shrinking and overflowing of lava, from a fluctuating pressure from beneath, would gradually produce the existing forms. We have nothing on a corresponding scale on the Earth; but the craters of the Sandwich Islands, Kirauea and Haleakala, the one a fused, the other a frozen lake of lava, with the small " blowing-cones " which eject only cinders and ashes, afford a striking analogy; and we can hardly wonder at some remaining difficulties, while geologists are so little agreed about " elevation-craters " and submarine volcanos on the Earth. The circular cavities of the Moon are arranged in three classes, — *Walled* (or *Bulwark*) *Plains*, *Ring-Mountains*, and *Craters:* a fourth includes little pits without a visible ring. The second and third differ only in size; but the first have a character of their own;—the perfect resemblance of their interiors with the grey plains, as though a great entrenchment had been thrown up around an undisturbed surface; and this offers much difficulty in the absence of atmospheric degradation. The idea of a corresponding covering of vegetation may cross the mind; but how maintained without air and water? It has been ingeniously suggested that a low stratum of carbonic acid gas—the frequent

E 2

product of volcanos—may in such situations support the life of
some kind of plants : and the idea deserves to be borne in
mind in studying the changes of relative brightness in some
of these spots.* The deeper are the more concave craters;
but the bottom is often flat, sometimes convex ; and frequently
shews subsequent disturbance, in ridges, hillocks, minute
craters, or more generally, as the last effort of the eruption,
central hills of various heights, but seldom rivalling that of
the wall. The ring is usually steepest within, and often built
up in vast terraces. Nasmyth refers these—not very probably—
to successively decreasing explosions ; in other cases he more
reasonably ascribes them to the slipping down of materials
upheaved too steeply to stand, and undermined by lava at their
base, leaving visible breaches in the wall above : they would
be well explained on the supposition just mentioned of fluctuating levels in molten lakes.   Small transverse ridges occasionally descend from the ring—chiefly on the outside ; great peaks
often spring up like towers upon the wall ; gateways at times
break through the rampart, and in some cases are multiplied
till the remaining piers of wall resemble the stones of a gigantic Druidical temple.   A succession of eruptions may be
constantly traced, in the repeated encroachment of circles on
each other ; the largest are thus pointed out as the oldest
craters, and the gradual decay of the explosive force, like
that of many terrestrial volcanos, becomes unquestionable.
The peculiar whiteness of the smaller craters may indicate

* Gruithuisen thought he perceived, in the grey tints of depressed
surfaces, some of which vary with the amount of solar light, traces of
several kinds of vegetation, comprised between 65° N. and 55° S. latitude, and preserving the correspondence observed on the Earth between
increasing latitude and elevation.

something analogous to the difference between the earlier and later lavas of the Earth, or to the decomposition caused, as at Teneriffe, by acid vapours.

4. *Valleys*, of ordinary character, are not infrequent; some of grand dimensions; others mere contracted gorges. The most curious of these are entitled to be classed apart as

5. *Canals*, (or *Rills*.) These were discovered by Schröter : Gruithuisen and Lohrmann added to their number, which Beer and Mädler raised to 90 in their work on the Moon ; but the latter astronomer, on taking charge of the noble Dorpat achromatic of $9\frac{1}{2}$ inches aperture, perceived more than 150, and thought it might be possible to descry 1,000 of these most singular furrows, which pass chiefly through levels, intersect craters (proving a more recent origin), re-appear beyond obstructing mountains, as though carried through by a tunnel, and commence and terminate with little reference to any conspicuous feature of the neighbourhood. The idea of an artificial origin* is negatived by their magnitude; they have been more probably referred to cracks in a cooling surface : but the observations of Kunowsky, confirmed by Mädler at Dorpat, seem in some instances to point to a less intelligible origin in rows of minute contiguous craters.

Wonders are here in abundance for the student : but he will find it impossible to pursue them far from the terminator, — they must be viewed under the oblique rays of a rising or

* Gruithuisen, following out his strange theory that the Moon was once surrounded by an immense ocean which has wholly disappeared (whither ?) considered the larger of these furrows as the beds of dried-up rivers; the smaller he referred to artificial clearings in the forests, answering the purpose of roads. It is to be regretted that the extravagance of his fancy should have brought discredit upon the unquestionable precision of his sight.

setting Sun. As the angle of illumination increases, a fresh aspect of things creeps in and successively extends itself over the whole disc, and in its progress the inexperienced observer will find himself astonished at the change, and frequently bewildered in the attempt to trace out the land-marks of the surface. Objects well recognised while the relief of light and shade remains will become confused by a novel effect of local illumination, and the eye will wander over a wilderness of streaks and specks of light, and spots and clouds of darkness, where it may sometimes catch the whole, sometimes a portion, sometimes nothing, of many a familiar feature; while unknown configurations will stand boldly out, defying all scrutiny, and keeping their post immoveably till the decreasing angle of illumination warns them to withdraw. Nothing can be more perplexing than this optical metamorphosis, so complete in parts as utterly to efface well-defined objects; so capricious as in some instances to obliterate one, and leave unaffected the other, of two similar and adjacent forms. Gruithuisen, carrying out an idea of Schröter's, referred some of these changes to the progress of vegetation, which, if existing, will naturally, in default of a change of seasons, run its whole course in a single lunation: even the cautious Beer and Mädler have admitted that some variations of colour may possibly point in this direction; and photographic results seem to indicate the presence of green light not cognisable by the eye: but the general change demands a more universal solution; and probably a wide range of colours in the soil may be concerned in the effect. The subject calls for careful study, but would involve much laborious application.

The most obvious features of the Moon under a high illumination are the *Systems of Bright Streaks* which issue, though in widely differing proportions, from seven different centres : all craters, few inconsiderable, but none of the very largest class. In some cases the streaks proceed from a circular grey border surrounding the crater; in others they cross irregularly at its centre. They pass alike over mountain and valley, and even through the rings and cavities of craters, and seem to defy all scrutiny. Nichol asserts that in some contiguous systems, the order of formation may be detected from the mode of their intersection ; * a statement well deserving the notice of those whose telescopes will carry them through the enquiry. But one thing is certain, that they cause no visible deviation in the superficial level ; a fact irreconcileable with Nasmyth's conjecture, that they are cracks diverging from a central explosion, filled up with molten matter from beneath; trap-dykes on the Earth are indeed apt to assume the form of the surface, but the chances against so universal and exact a restoration of level, all along such multiplied and most irregular lines of exposure, would be incalculable: many of the rays are also far too long for this supposition, or for that of Beer and Mädler, that they may be stains arising from highly heated subterraneous vapour on its way to the point of its escape. The extraordinary brilliancy of some portions of the Full Moon is less difficult of explanation, when we bear in mind the effect of chalky strata, or that peculiar kind of granite which on the loftiest peaks of the Himalaya may readily be confounded with the adjacent snow. In many cases we have no key to this vivid reflection in the form of the surface ; in others it occupies a marked position, as on the summits of rings and central mountains: the idea of a mirror-like glaze, reflecting an image of the Sun,

* The following, according to him, is the chronological order of three great systems: *Copernicus, Aristarchus, Kepler.*

though entertained by such authorities as Beer and Mädler, seems difficult to be reconciled with the ever-varying angle of illumination.

A few *peculiarities of arrangement* deserve to be mentioned here. The remarkable tendency to circular forms, even where explosive action seems not to have been concerned, as in the bays of the so-called seas, is very obvious; and so are the lateral lines of communication already mentioned. The gigantic craters or walled-plains often affect a meridianal arrangement; three huge rows of this kind are very conspicuous, near the centre, and each of the limbs. A tendency to parallel direction has often a curious influence on the position of smaller objects; in many regions these chiefly point to the same quarter, usually N. and S., or NE. and SW.; thus in one vicinity Beer and Mädler speak of 30 objects following a parallel arrangement, for one turned any other way; even small craters entangled in such general pressures have been squeezed into an oval form; and the effect is like that of an oblique strain upon the pattern of a loosely-woven fabric : an instance of double parallelism, like that of a net, is mentioned, with crossing lines from SSW. and SE. Local repetitions frequently occur; one region is characterised by exaggerated central hills of craters; another is without them; in another, the walls themselves fail. Incomplete rings are much more common towards the N. than the S. pole; the defect is usually in the N., seldom in the W. part of the circle; sometimes a cluster of craters are all breached on the same side.* Two similar craters often lie N. and S. of each other, and near them is frequently a

---

* Compare Elwes's account of the small cones on the floor of the great extinct crater Haleakala, some of them " broken down at the side, nearly always on the NE."—Sketcher's Tour, 214.

corresponding duplicate. Two large craters occasionally lie
N. and S., of greatly resembling character, S. usually ¾ of
size of N., from 18 to 36 miles apart, and connected by ridges
pointing in a SW. direction. Several of these arrangements
are the more remarkable, as we know of nothing similar on
the Earth.

The question as to the *continuance of explosive action* on the
Moon is one of great interest. It is now generally understood
that the volcanos, which Herschel I. and others thought they
saw in eruption on the dark side, were only the brighter spots
reflecting back to us the earth-shine of the lunar night with
the same proportional vivacity as the sunshine of the day. No
valid reason indeed has been assigned for the fact, witnessed by
many observers, especially Smyth, that one at least of these
spots,—Aristarchus,—varies remarkably in nightly luminosity
at different periods,* nor for the specks of light which more
than once Schröter caught sight of on the dark side for a short
time : † but in these cases there has been no subsequent perceptible alteration of surface ; and they furnish no reply to the
enquiry respecting present changes. Such were abundantly
recorded by Schröter in his day, chiefly variations in the visibility or form of minute objects ; but a great majority of them
he, and subsequently Gruithuisen, who witnessed many such
appearances, referred to the lunar atmosphere ; and Beer and
Mädler are disposed to discard them all as the result of inac-

* Schröter's conjecture that such variations, which he observed in several parts of the disc, may be due to atmospheric condensation during
the lunar night, is more elegant than probable, yet may deserve consideration.

† Such a phænomenon, more extended, but very faint, was seen, or
fancied, also by Gruithuisen.

curacy or varying illumination. The extraordinary influence
of the latter upon the aspect of distant and unknown objects
may be estimated by any one who will sketch on paper the
changed effects of light and shade on any familiar terrestrial
object at different times of day; still there seems to be a
residuum of minute variations not thus disposed of, and in
some cases possibly indicating actual change of surface: nor
is there any improbability in the idea that, as on the Earth,
disturbing agency may have greatly diminished without having
become extinct: at any rate observation, not assertion, must
decide the point. There are no traces of any great convulsion
since the date of the first lunar map; but as to minuter
changes, we must recollect that we have not till lately possessed
the means of detecting them. Schröter's drawings are very
rough; the much more careful ones of Lohrmann, and Beer
and Mädler, are too recent to warrant great expectations:
another half-century may fully prove the continuance, or infer
the extinction, of that once mighty transforming force: in the
meantime some observations of my own indicate its existence,
though to a very limited extent. Those who feel interested in
such details should procure the beautiful and not expensive
chart of Beer and Mädler, to which the explanatory work
"Der Mond" (The Moon) is an important addition for the
German scholar: their labours ought to be more generally
known in this country.

Little of a satisfactory nature can be said as to a *Lunar
Atmosphere*. That it must be far rarer than any known gas is
demonstrable from theory, and proved by observation, which
shews us a sharply defined outline, and detects no refraction in
stars over which the Moon passes: and hence its entire absence
has been maintained by great astronomers: Schröter asserted

its existence from many changes of aspect in minute objects, and from a very dim twilight which he traced through 9 years beyond the points of the horns; his inferences are supported and in part exceeded by Gruithuisen, who frequently saw — or imagined — fogs and clouds on the surface. Schröter explains the defined limb and the absence of refraction by limiting the atmosphere to the inferior regions, and leaving the higher grounds free. Beer and Mädler, while explaining away — not very satisfactorily, — Schröter's twilight, which they could never distinctly find, do not deny the possibility of a very rarefied gaseous envelope.* Those traces of twilight which Gruithuisen confirms, and which I imagined I saw, 1855, June 20, but could only speak doubtfully from want of better optical means, might well engage the student's attention; in order to assist him two figures are here given from among Schröter's numerous delineations. They represent the cusps of the Moon, as seen by him, 1792, Feb. 24, with a 7 feet reflector by Herschel I., powers 74 and 161.

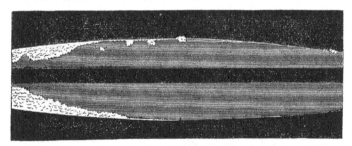

Occultations of planets or stars, which are important to the professed astronomer, are interesting to the amateur as bringing

---

* Some very curious photographic experiments by De La Rue seem to tend the same way; but these researches are as yet incomplete.

out distances and motions of which the naked eye takes little
cognizance: the obvious difference of focus for the star and
Moon demonstrates at once the comparative nearness of the
latter; and its orbital movement is made apparent from minute
to minute. A grand effect is produced by the visible sailing
of this ponderous globe through immeasurable space; and it
may well convey a deep impression of the omnipotent Power
and consummate Wisdom which orders its undeviating course.
The instantaneous extinction, too, and sudden flashing out of
a large star in these circumstances are very striking, but less
instructive than might have been expected as to the question
of an atmosphere: the immersion and emersion are usually as
sudden as if none existed; but this may be accounted for by
the rapid motion of the Moon, or on Schröter's theory, by the
elevation of that portion of the limb: at other times, a gradual
extinction may be only due to irregularities on the edge: the
distortions or flattenings occasionally noticed in the shape of
planets are too extensive for any such cause, and must have
some optical origin. All occultations of importance are predicted in the Nautical Almanac, or that of Dietrichsen and
Hannay, an excellent popular substitute: the most beautiful
are *immersions* behind the *dark* limb when the latter receives
a strong earth-light, as between New Moon and I. Quarter.*

The *projection* of a star upon the Moon, just before immersion, as if it advanced in front of it, is very difficult of
explanation. South, after a careful examination of the instances on record, was unable to come to any satisfactory
conclusion: it is an occurrence of the most capricious kind,

* Kunowsky has seen the dark limb sharply defined 3ᵈ 4ʰ after I.
Quarter. Gruithuisen professed that he was able to trace the larger
seas on the dark side with the naked eye.

and can never be predicted for any star, eye, or telescope: optics, and not astronomy, are evidently concerned in it.

Libration must be well understood, before proceeding to topography. This is an apparent displacement of the spots with respect to the limb, arising from the equable rotation of the Moon on its axis combined with its unequable velocity in an elliptical orbit, as well as from a slight inclination of the axis to the plane of the orbit: it completes its changes in about four weeks, though an exact restoration of the position called "mean (or medium) libration" does not take place till the end of three years. Hence, though we are commonly said to see always the same hemisphere, it is only approximately true. The spots are constantly swinging a little backwards and forwards, and those near the E. and W. limbs going alternately out of sight, while, as our eye rises successively above each pole, a little more or less of those regions is seen,— the whole area thus concealed and exposed by turns amounting to $\frac{1}{7}$ of the surface of the globe. No map of the Moon, therefore, can correctly represent its whole aspect on two following nights, and they are of course constructed to correspond with a "mean," or average state of libration. The displacement of the spots, which amounts at a maximum to 6° 47′ N. and S., 7° 55′ E. and W., and 10° 24′ from a concurrence of both, produces no great effect upon the aspect of the central parts, beyond a slight shifting with regard to the Moon's equator or 1st meridian; but in the foreshortened neighbourhood of the limb its results are very obvious. Fresh regions constantly take the outline of the disc, and the mountainous projections of to-night may be all out of sight to-morrow. One effect of libration is, that the spots have no fixed position with respect to the Moon's age, being sometimes

earlier, sometimes later on the terminator, so that no precise
instructions can be given when to look for them. A few
approximations alone will therefore be found in these pages,
especially as a little practice with the map will render identification easy : the following considerations, however, may be
useful to the beginner.

The relief of the surface will be stronger as it is nearer
to the terminator, and all delicate and difficult objects are
best seen near the sunrise or sunset of the Moon, which,
like the corresponding times on the Earth, abound with grand
and beautiful effects of light and shade. Every little irregularity then assumes temporary importance ; inconsiderable
hillocks, minute craters, low banks, narrow canals, become
visible in the horizontal ray ; rising grounds or soft valleys
seem to start into existence, the outer slopes of craters advance
into the surrounding levels, larger masses appear in exaggerated prominence,

"Majoresque cadunt de montibus umbræ."

These broad and deep shadows often taper off to such slender
points, that a caution may be requisite, not to infer anything
extravagant respecting the sharpness of the form which casts
them. It may be questioned whether anything in the Moon
exceeds the acuteness of the Swiss Finsteraarhorn, or the
abruptness of the Pic du Midi d'Ossau in the Pyrenees. An
oblique light cast upon a rough surface, modelled in clay or
dough, will show how disproportioned oftentimes is the
shadow to the reality ; and the same experiment will prove to
us that in many cases the true relief will be unknown till the
shadow on each side of the object has been examined. The

intense blackness of the lunar shadows gives an effect which
must be strangely contrasted with a distant prospect of our
Earth. Here, a highly reflective and refractive atmosphere
surrounds all objects with a constant illumination, which is
continued even after sunset : there, the Sun goes up and
down in noon-day strength, and the shadows are unrelieved
by any reflection from a sky which must be almost black all
day : every mountain produces utter darkness where it inter-
cepts the Sun, and every crater, while its ring is glittering like
snow in the rising or setting beam, is filled with midnight
shade. Dawes, alone, by employing the contracted field of
his solar eye-piece, has traced a very faint glimmering in
these depths, produced by reflection, not from the atmosphere,
but the cliffs above, illuminated by the full sunshine. The
terminator is indeed marked through the grey plains by a
narrow shadowy border, and the tops of mountains just
appearing or vanishing in the night-side are somewhat defi-
cient in brightness; but this is the " penumbra," or partial
illumination, due to a portion of the Sun's disc, while the rest
is beneath the horizon. In other respects the glittering
sharpness of the Moon's sunrise, or sunset, is widely contrasted with the softness of our own skies. But though it is
then that we must watch for minute details, for large objects
we shall find the relief too broad and partial ; unbroken night
conceals the greater cavities, and the shade of loftier summits
renders lower ranges invisible. Not till day descends among
the terraces of the one, or creeps down over the shoulders and
along the slopes of the other, does their true structure come
out; and their distinctness increases while their minuter
neighbours decline into insignificance.

Lohrmann,* and Beer and Mädler, have done admirably well in their delineations. Yet a little experience will shew that they have not represented all that may sometimes be seen with a good common telescope. My own very limited opportunities have satisfied me, not only how much remains to be done, but how much a little willing perseverance might do, provided there were some knowledge of the laws of perspective and shadow, and a due attention to the direction of the incident and reflected light: some correctness in design would also be very desirable, and if proportion is tolerably observed, a number of rough sketches under varied lights would be more serviceable than one or two finished drawings. Detail being the great object, a small portion only should be attempted at once; this will not merely be easier and more pleasant, but will avoid the change in the shadows, which is considerable during a long-continued delineation. The record, to be of value, must possess four data: 1. *Hour of Observation*;—2. *Moon's Age*, reckoned from or to the nearest change;—3. *Position of Terminator* referred to any adjacent well-marked spots;—4. *Libration*, indicated by time reckoned to or from nearest epoch of greatest libration, as given in Nautical or Dietrichsen's Almanac. More than this cannot be ascertained without a micrometer; but this is enough for the comparison of observations. If the last three data are nearly coincident, the angles under which the landscape is illuminated and viewed differ so little in the two observations, that any variation in its details must be either referred to,—1. inadvertence or mistake in the observer; —2. actual change in the Moon's surface; or—3. obscuration or deception in her atmosphere. Schröter relates so many

* The discontinuance of this great observer's work, in consequence of his failing sight, has been a serious injury to selenography.

instances of the latter kind that, even if we reject the confirmation of the keen-eyed but fanciful Gruithuisen, it seems
difficult to follow Beer and Mädler in disposing of them all
as errors of observation ; nor do I believe that the enquiry is
unworthy of notice. Gruithuisen gives one important caution,
which Schröter probably sometimes neglected ; when objects
are lengthened due E. and W., a slight amount of N. or S.
libration, by causing the light and shade to change sides, may
widely vary their appearance. Little or nothing is gained by
very high magnifiers : Beer and Mädler did not exceed 300 ;
Schröter commonly used a lower power.

A good popular Map of the Moon has been hitherto a desi-
deratum in England. Russell's and Blunt's, striking as to general effect, break down in details. Those in ordinary books of
astronomy are still more useless. The one here given will, it
is hoped, be found less defective : it makes no claim to pictorial
resemblance, and professes to be merely a guide to such of the
more interesting features as common telescopes will reach. It
is carefully reduced from the " Mappa Selenographica" of Beer
and Mädler, published in four sheets in 1834, on a scale of
3 feet 1¼ inch, omitting an immense mass of detail accumulated
by their wonderful perseverance,* which would only serve to
perplex the beginner. Selection was difficult in such a crowd;
on the whole it seemed best to include every object distinguished by an *independent name;* many of little interest thus
creep in, and many sufficiently remarkable ones drop out ; but
the line must have been drawn somewhere, and perhaps would
have been nowhere better for the student. Other spots, however,

* Some idea may be formed of this, from the 919 micrometrical measures of the positions of spots, and 1,095 of heights and depths, contained
in their work " Der Mond."

F

have been admitted from their conspicuousness, to which
Beer and Mädler have given only a *subordinate* name;
minuter details come in, in places, for ready identification;
elsewhere, larger objects are passed by, as less useful for the
purpose of the map. The nomenclature is that established by
Beer and Mädler. Hevel (or Hevelius), in the earliest attempt, designated different regions of the Moon from supposed
geographical analogies; but this system has long been abandoned, except in the case of some mountain ranges. Riccioli,
a far inferior observer, adopted a more available method of
affixing to the larger spots the names of distinguished philosophers; his list was increased by Schröter, who made each
name include the adjacent objects, by adding the letters of the
alphabet; and this system, improved and generalised, has been
applied by Beer and Mädler to the whole disc; the name
used alone, distinguishes the principal object; Roman or
Greek letters added to it signify respectively the elevations or
hollows in the vicinity. To avoid crowding our map, letters
and numbers are substituted for names; every object in the
descriptive notices will be thus referred to, but as this selection is limited, a complete list of names is subjoined, which
will also secure, in all cases, the important object of identification as far as it goes.

The points of the *Lunar Compass* must be mastered before
we can use the map. Astronomers have fixed these from their
relative, not intrinsic, position: that is, the several portions
of the disc are named from the adjacent quarter of the sky
when the Moon is on the meridian. Hence, N. and S. occupy
the top and bottom, but E. and W. are reversed, as compared
with terrestial maps; the former being to the left, the latter to
the right. But maps of the Moon usually represent its tele-

scopic, that is, inverted appearance, so that we shall find S. at top, N. at bottom, E. to right, W. to left. The meridians and parallels of latitude have been omitted in the map, except the 1st Meridian and Equator, which divide it, like the original, into Four Quadrants: these are called by Beer and Mädler, the 1st or NW., 2nd or NE., 3rd or SE., 4th or SW. Quadrant. A selection of the most interesting objects in each follows, from materials furnished by the "Mond" of these astronomers, and retaining their arrangement: additions are made from other sources; but all statements not otherwise authenticated depend upon their authority. In the more remarkable instances, their measures of height and depth are given; these were ascertained by the lengths of the shadows, a method previously employed with less accuracy by Schröter; capable of much precision on favourable ground, but elsewhere uncertain. For our present purpose round numbers will be a fully sufficient approximation.

First, or North-West Quadrant.

Mare Crisium (a on the map). We begin with a conspicuous dark plain, the most completely bounded on the Moon, and visible to the naked eye: apparently elliptical from foreshortening, but really oval the other way, being about 280 miles from N. to S., 354 from E. to W., and containing 14,260 square miles, or $\frac{1}{111}$th part of the visible hemisphere. Its grey hue has a trace of green in the Full; this has also been represented by the present talented Astronomer Royal for Scotland, C. Piazzi Smyth, in two beautiful figures taken during the increase and wane. On rare occasions it has been seen by Schröter, and in part by Beer and Mädler, speckled with minute dots and streaks of light: something of this kind I saw with a fluid achromatic, 1832, July 4, near I. Quarter; it would be difficult to say why, if these are permanent, they are so seldom visible. The surface is deeply depressed, lower than Mare Fœcunditatis and M. Tranquillitatis. The boundary mountains are in part very steep and lofty. The PROMONTORIUM AGARUM (1) rises about 11,000 feet; a mountain SE. of Picard 15,600 feet, rivalling our Mont Blanc. On W. edge, Schröter delineated a crater called by him ALHAZEN, which he employed to measure the existing libration : he saw in it after a time unaccountable changes, and now, it is said, it cannot be found. Beer and Mädler think he confounded it with a crater (2) lying further S.; the question, however, which in the interim was debated between Kunowsky and Köhler, is not quite cleared up. The plain contains some moderate-sized craters — the largest, PICARD (4), S. of which Gruithuisen saw some curious regular white ridges like walls ; — and several very minute ones. Near E. edge, where there is a kind of pass in the great surrounding ridge, lie several small, but in part lofty mountains,*—islands, as it were ;—among these Schröter describes singular changes, which he refers to an atmosphere ; Beer and Mädler consider them merely varied illumination, or pass them by as unworthy of attention. Succeeding observers may not feel satisfied with this summary decision. Schröter was a bad draughtsman, used an inferior measuring apparatus, and now and then made considerable mistakes; but I have never closed the simple and candid record of his most zealous labours with any feeling approaching to contempt ; and though there may be truth in the assertion of Beer and Mädler, that he was biassed by the desire of discovering

* These are ill-figured in the great map, and Beer and Mädler have given a separate representation of them in their " Mond."

changes, they, possibly, were not themselves free from an opposite prepossession.

Central hills are absent from the craters of this district.

M. Crisium may be well seen about 5^d after New, or 3^d after Full; in the latter case it is a magnificent spectacle when intersected by the terminator and partially covered in by the vast shadows of the mountains, from which Schröter considered that those on NE. side must be at least 16,000 or 17,000 feet high.

This astronomer has inserted in his work a marvellous observation by Eysenhard, a pupil of Lambert, 1774, July 25. The night being perfectly clear, he saw with a common 4 feet refractor, four bright spots in M. Crisium, then intersected by the terminator, two of which only,—those on the day-side, —can be identified; after noticing them at times for 2^h, he found all at once that the part of the terminator in M. Crisium had a slow reciprocating motion, completed in 5 or 6^m, between these pairs of spots, each pair being touched by it in turn. Two other refractors of 7 and 12 feet shewed this appearance with equal distinctness, and it was observed for 2^h, the terminator in M. Fœcunditatis remaining perfectly still. This is a very strange story; yet Lambert seems to have believed it; and perhaps we cannot pronounce it wholly incredible in the face of an equally wonderful and perfectly well-attested retrogression in a satellite of Jupiter, to be described hereafter.

Between FIRMICUS (7) and the limb, about half-way in mean libration, are some curved dark streaks, in which Beer and Mädler found singular variations, resulting, as they admit, possibly (not probably) from periodical vegetation.

CLEOMEDES (12), a walled plain 78 miles in diameter, includes a small crater (Cleomedes A), brilliant, but not always alike defined, in Full; Schröter had found it not always equally visible. He speaks of many variations in the interior level, and represents it rather differently from Beer and Mädler. Gruithuisen found its W. part marked out into many rhomboids, — squares in perspective.

BURCKHARDT (**19**), 35 miles in diameter, lies 12,700 feet below its E. wall.

GEMINUS (**20**), 54 miles broad, has a ring 12,300* feet high on E., 16,700 on W. side.

BERNOUILLI (**21**), equally deep, is very precipitous.

GAUSS (**22**) is a walled plain, 110 miles long. Beer and Mädler describe the fine effect of sunset upon its ring. It has a grand central mountain, which must at times command a glorious view, across a plain of 50 miles covered with night, to illuminated peaks all round the horizon, above which the Sun on one side, and the Earth on the other, are slowly coming into sight.

STRUVE (**25**), a slight depression, is remarkably dark in Full.

ENDYMION (**27**), a walled plain 78 miles in diameter, is, in some states of libration, very dark in Full: the irregular wall rises W. to more than 15,000 feet, overtopping all but the very highest peaks of our Alps. I have seen it in grand relief 3ᵈ 7ʰ after New, 2ᵈ 9ʰ after Full. Between **27**, **28**, and **29**, is a curious double parallelism, objects all ranging SSW. or SE.

---

* By a singular coincidence, in reducing as usual French to English measure, the resulting number was 1 2 3 4 5, with 6 in the first place of decimals. The chances against such a sequence must have been extremely great: but it exemplifies a principle, not always kept in mind, that so long as a thing is *possible*, it must *sometimes* occur.

ATLAS (**28**). A superb amphitheatre 55 miles broad, 460 square miles in area; its ring rich in terraces and towers, rising 11,000 feet on N.; a very dark speck in interior.

HERCULES (**29**). A worthy companion to it, 46 miles across. The ring, in places double, includes a small crater of subsequent date. Look for this pair 5 or 6^d after New, 3$\frac{1}{2}$^{d} after Full.

FRANKLIN (**32**). Several incomplete rings lie hereabout, all open N.

MARE HUMBOLDTIANUM (**B**), discovered by Beer and Mädler,* is rather more than half as large as M. Crisium, but close to the limb, above which the peaks of its W. border sometimes appear in profile, at least 16,000 feet high.

MOUNT TAURUS (**51**) is a lofty range, containing the terraced crater RŒMER (**52**), 26 miles wide, 11,600 feet deep.

POSIDONIUS (**54**). This walled plain, nearly 62 miles across, includes several small objects, in which Schröter found repeated changes: the shadow in the bright little crater varied in length, and was once replaced by a grey veil. Beer and Mädler never saw anything unusual. A good object about 6^d after New.

LITTROW (**55**) and MARALDI (**56**) shew the SSW. parallelism of the vicinity: VITRUVIUS (**57**), with a very dark interior, lies in a mottled region, in one place slightly tinged with blue.

---

* Part, however, is shown on Russell's Lunar Globe (1797), a beautiful work of art, but faulty in detail. Madame Witte, a Hanoverian lady, has executed a most perfect globe in relief, from Beer and Mädler's observations and her own. Sir Chr. Wren, when Savilian Professor at Oxford, made a lunar globe in relief at the request of the Royal Society and by command of Charles II., who placed it in his cabinet. It is to be hoped that it has been carefully preserved.

[MOUNT ARGÆUS] (**58**). A small range gradually rising towards E. to a summit, perhaps like the Niesen above the lake of Thun, but much loftier, and ranked by Schröter as high as the loftiest of our Pyrenees. It is remarkable for the spire of shade which it casts across the plain at sunrise; but it has been unjustly treated by Beer and Mädler, being omitted from their nomenclature, which contains many very inferior hills; as it is a very interesting object, I hope I am not guilty of presumption in proposing a name for it, chosen from its vicinity to Mount Taurus. It requires close watching, as the shadow rapidly loses its slender point; look for it when the ring of Plinius just heaves in sight beyond the terminator. I have seen it thus 4^d 21^h after New.

MACROBIUS (**59**), nearly 42 miles in diameter, is almost 13,000 feet deep.

PROCLUS (**60**) has a ring, next to **148** the most luminous part of the Moon; yet scarcely ever visible on the dark side; very much less so than many duller objects. Beer and Mädler ascribe this to the narrowness of the wall. It is the centre of several bright streaks, not very easily seen.

PALUS SOMNII (**F**), an uneven, defined, always distinguishable surface, has a peculiar tint, perhaps yellowish brown, unlike the simple grey of MARE TRANQUILLITATIS (**G**).

PLINIUS (**61**). A terraced ring 32 miles broad, filled with hillocks.

MENELAUS (**70**). A very steep crater about 6,600 feet deep; the ring is very brilliant in Full. All the ridges running from it trend SW., and this parallelism reaches over a great district, including most of the Apennines, Hæmus, M. Vaporum, and the neighbourhood. **70** is the centre of several bright streaks—one is a continuation of a ray from **180**, ex-

tending in all more than 1,850 miles. This spot was one of
Herschel I.'s pseudo-volcanos.

MARE SERENITATIS (**H**). This wide and beautiful plain is
nearly circular, about 432 miles from N. to S., by 423 from
E. to W.; the interior appeared at Full to Beer and Mädler of
a clear light green, unnoticed by Schröter and Lohrmann,
and not easily seen even by its discoverers; it is set in a
border of dark grey, and bisected by a straight whitish streak,
invisible, like all of its class, near the terminator. Towards
the W. edge of the plain is a curious long low serpentine
ridge, discovered and figured by Schröter, but only well seen
from its shadow near the terminator. I have found it thus
between 5 and 6^d after New.

CAUCASUS (**75**). This grand mountain mass rises into insulated peaks, suitably termed "aiguilles" by Beer and Mädler,
as lofty as any on the Moon, those on the limbs excepted, and
reaching 18,000 or 19,000 feet. Their narrow shadows are
drawn out into fine points—a superb spectacle, which I have
seen about I. Quarter.

EUDOXUS (**77**) and ARISTOTELES (**78**). A noble pair of
craters, not easily found in Full, in a region then dotted with
literally thousands of bright specks. The terraced wall of
77, 11,300 feet above the W. interior, is on that side crowned
by two turrets of 15,000 feet. **78** is more than 50 miles
broad, nearly as deep as **77**, but with a much richer wall.
The interior of both resembles the ring more than is usual.
78 is very remarkable for rows of minute hillocks lying NE.,
NW., and SW., in lines pointing to the centre of the crater;
the finest specimen of this not uncommon arrangement : it is
less distinct round **77**. It requires very favourable light, and
is difficult for the student. I have seen it, with wide SW.

libration, about I. Quarter. S. and SW. of **77** the surface down to M. Seren. is thronged with hillocks innumerable, like stars in the most crowded part of the Galaxy: on E. they reach to Caucasus.

WEDGE-SHAPED VALLEY OF ALPS. An extraordinary cleft, figured by Bianchini in his work on Venus; a circumstance unnoticed alike by Schröter, and Beer and Mädler. About 83 miles long, and from $3\frac{1}{2}$ to $5\frac{3}{4}$ miles broad, it breaks in a straight line through the very loftiest Alps, with precipitous sides and a depth of at least 11,500 feet, in which Mont Perdu or the Vignemale of the Pyrenees would disappear. It is always easily found. Between this valley, **79** and **I**, lies a curious region all sprinkled with hillocks: Schröter estimated them at 50; Beer and Mädler at 700 or 800 at least; they reach 920 miles in length by 90 to 230 in breadth. I once saw them finely developed at I. Quarter, in the 20 feet achromatic of $14\frac{3}{4}$ inches aperture, exhibited by Slater the optician.

ALPS (**80**). A lofty and exceedingly steep chain, rising into separate peaks: one of the highest, the Mont Blanc of Schröter (next to which our number stands) reaches, he says, more than 13,000 feet. Beer and Mädler reduce it below 12,000. Close beneath its E. foot, Schröter perceived, 1788, Sept. 26, on the dark side of the Moon, a small speck of light, like a 5 mag. star to the naked eye, which having been verified in position, and kept in view for fully 15ᵐ disappeared irrecoverably :

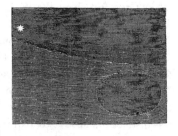

in or near its place he subsequently found in the day-side a round black shadow as of a deep hollow, which at other times, though under similar illumination, looked only grey. A copy of his figure is here given.

PALUS NEBULARUM (ɪ) and PALUS PUTREDINIS (ᴋ) are generally so level, that mounds of 50 or 60 feet would be rendered visible by their shadows near the terminator.

CASSINI (**81**). A curious ring-plain, whose narrow wall encloses a scarcely depressed space, and contains a small deep crater. It casts long spires of shade about I. Quarter.

THEÆTETUS (**82**) would take in Snowdon twice over, with still depth to spare.

ARISTILLUS (**83**) is a very grand crater, 34 miles broad, 11,000 feet deep on E. side, with a fine central mountain. The ring stands up nobly from the plain, and is flanked on all sides (as Schröter perceived) by radiating banks resembling lava-streams. I have seen them well about 1ᵈ after I. Quarter.

AUTOLYCUS (**84**), its smaller companion, is nearly as deep, with a similar but less evident flow of lava. This pair of craters illustrates a not uncommon arrangement, of two rings in the same meridian; very similar in every respect; if unequal, in diameter as 3 to 4, and the smaller lying S.; from 18 to 36 miles apart; and connected by low ridges running SW.

APENNINES (**85**). This very extensive chain is more like the mountains of the Earth than is usually the case. Its length, according to Schröter, is nearly 460 miles, great part of it lying in the next Quadrant. Its SW. side ascends gradually — the opposite aspect breaks down at once in awful precipices of stupendous height, casting a shade through 83 miles, or losing the last ray as far beyond the terminator. Huygens (**90**), the loftiest peak, rises, according to Schröter, no bad measurer, to 21,000 feet. Beer and Mädler think he confounded two adjacent summits, and give 18,000 feet: in either case a superb elevation, greatly overtopping our Mont Blanc, perhaps rivalling Chimborazo. On its apex is a minute crater. Other summits, though inferior, are yet of extraordinary height and steepness: Hadley (**87**), facing the Caucasus, 15,000 feet, — Bradley (**89**) 13,000 feet, — Wolf (**92**) 11,000 feet. The gradual entrance of the range into sunshine about I. Quarter is a glorious spectacle; and its projection into the dark side, which may be seen by a keen eye, probably gave rise to the early idea mentioned by Plutarch, that the Moon was mountainous: from its abruptness it does not lose all traces of shade till 24^h before Full. It contains very few craters, and consists of ridges and peaks not to be numbered by thousands in large instruments: Schröter detected this; Mädler says it would take the opportunities of three or four years to delineate all the details which a power of 800 or 1,000 will shew.

CANAL OF HYGINUS. A fine specimen of these furrows lies fortunately in an excellent position in the MARE VAPORUM (**I**), a tract, of which the dusky streakiness has often interested me about I. Quarter. The canal is conspicuous enough to yield to a power of 40 in a fine telescope, under any illumination. It begins at the foot of a long low hill, as a flat valley about 1½ mile wide by 9 miles long, which contracts to about ¾ mile, with precipitous sides and great proportionate depth : it passes thus by four minute craters, and traverses HYGINUS (**93**) in a very singular manner, cutting down the ring, and in crossing the interior throwing up banks on either hand, which were once seen by Beer and Mädler as narrow bright lines, while the crater-bottom lay yet in shade. It bends a little here, and after receiving a narrow lateral branch, and touching on five very small craters and two broad hills, becomes.wider and shallower, and ends after about 106 miles much as it began. S. of this canal are two small dark spots, one larger, and one of a greenish hue: on the N. side is a region worthy of notice for its change of colour: in Full it is tolerably luminous: at Quadratures, there is near Hyginus a large blackish spot, covering two mountains and the vale between. Here Beer and Mädler, most unprejudiced, if not oppositely-prejudiced witnesses, admit a variation of colour not dependent merely on the angle of reflection, and possibly connected with changes like those of our seasons.

CANAL OF ARIADÆUS. This, discovered like the preceding by Schröter, lies W. of it; it is longer, broader, and probably deeper. It seems to pass under the first two mountains in its course, but the others it cleaves visibly, though greatly narrowed by one lofty hill. The accurate Lohrmann saw it passing through them all, and traced it for 175 miles. Gruithuisen and Kunowsky alone have detected a minute prolongation of it far through M. Tranquill. Both these canals may be looked for about I. Quarter.

TRIESNECKER (94) is surrounded by several minute canals intersecting and uniting with each other in a manner of which there is no other known instance.

MANILIUS (95), a beautiful cavity, 25 miles in diameter and 7,700 feet deep, has a broad luminous terraced crater- and peak-besprinkled ring. Being visible on the night-side, it was one of Herschel I.'s imaginary volcanos. Schröter has seen variations, not easily accounted for, in the relative visibility of this and **70** in the earth-shine.

JULIUS CÆSAR (96) and BOSCOVICH (98) are very dark hollows, DIONYSIUS (**99**) and SILBERSCHLAG (**101**) very brilliant crater-rings.

Near the two fine craters AGRIPPA (**102**) and GODIN (**103**) (which exemplify the remark under **84**), Schröter once saw for a short time on the dark side, a minute point of light.

RHÆTICUS (**104**), an irregular crater, marks exactly the Moon's Equator, and is one of the few spots to which both the Sun and Earth may be vertical. In its interior Gruithuisen detected one of those regular formations of which he said so much: his figure in the Astronomische Jahrbuch for 1828 represents a somewhat curved line with a little crater at one end and a mound at the other, crossed by four shorter straight lines; as he blamed its inaccuracy, it is not worth copying, but the object should be looked for, especially as its site is so convenient: a year afterwards he complained that it had since been greatly obscured "selenospherically," and seldom visible.

Second, or North-East Quadrant.

SCHRŒTER (**106**). This imperfect crater in an intricate district where the levels are bright, the hills and valleys very dark grey, is the guide to a very remarkable spot a little N., the number on the map standing between the two objects. Here, in 1822, Gruithuisen discovered the regular formation which at first attracted so much attention, and subsequently, from the fanciful character of the observer, fell into unmerited oblivion. The object, which he called Schröter (a name transferred by Beer and Mädler, when they could not find it, to the nearest crater), and which he maintained, notwithstanding its size, to be a work of art, is, as he describes it, a collection of dark

gigantic walls, visible only close to the terminator, extending
about 23 miles in each direction, and arranged on each side of
a principal meridianal wall in the centre, from which they slope
off SE. and SW. respectively, at an angle of 45°, like the
ribs of an alder leaf; those to the W. abutting on a meridianal
side wall, beyond which Schwabe and himself subsequently
perceived their continuation : frequently however traces only
of the figure could be discerned, as though it were obscured by
clouds : from this cause he supposed it had been quite misre-
presented by the careful Lohrmann, though he had studied it
for six months: some years afterwards it appeared to have
undergone a sudden change, and Gruithuisen himself could
scarcely make it out: Beer and Mädler, notwithstanding great
pains, were quite unsuccessful, till in 1838, subsequently to
the date of their map, they surveyed and measured the region
with the great Berlin achromatic of $9\frac{1}{2}$ inches aperture ; then
at last, among a number of curious labyrinthine details, of
which their " Beiträge " contains a view, five parallel valleys
appeared side by side, each about 9 miles by $3\frac{1}{2}$, sloping all to
SW., with prolongations in the same direction, giving very
fairly half of Gruithuisen's figure—a remarkable attestation
from an unfavourable quarter. But he could call many earlier
witnesses, having shewn it to several German philosophers ;
Prince Metternich, too, at Vienna, found it from his description,
and it was beautifully seen by Schmidt and Schwabe. In
England I believe it has never been seen—possibly never
looked for. But we ought to look for it, and to see it,—unless
we prefer to admit the justice of Gruithuisen's taunt, " the
scientific John Bull here has gone empty away, and behaved
himself about it just as his natural disposition led him." Had
I met with Gruithuisen's works in time, I would have tried to

find it before these pages left my hands, but I heartily wish my readers success. No great optical power is needed, — he had apertures of only about 3 and 4¼ inches; nor can a beautifully keen sight like his be indispensable, as so many others in Germany saw it well. But much perseverance will be required to catch the proper moment. Though Gruithuisen considered it artificial, it does not appear that he thought it, what it has been called, a fortification.

SINUS ÆSTUUM (**N**). The absence of even the minutest crater here is unique. Mädler subsequently picked out a few with the Dorpat telescope.

ERATOSTHENES (**110**) and STADIUS (**111**), neighbours, the former more than 37 miles broad, the latter 5 miles broader, connected by a steep mountain of 4,500 feet (higher than any land in the British Isles) are curiously contrasted: **110** has a fine central hill, **111** is nearly level. **110** has a widely terraced wall of very irregular height; the surrounding ground is also very uneven; so that on E. side it is 16,000 feet above the inner, 7,500 feet above the outer surface; on W. only 10,000 feet and 3,300 feet. **111** has such an insignificant ring, about 130 feet high, that Beer and Mädler did not find it till after a search of three years. It seems, however, to have been seen in Riccioli's old refractor, and I found little difficulty with it. **110** is about as large as the county of Hereford, and lies intermediate betwen most dissimilar regions.

COPERNICUS (**112**). One of the grandest craters, 56 miles in diameter. It has a central mountain, two of whose six heads are conspicuous; and a noble ring composed not only of terraces, but distinct heights separated by ravines: the summit, a narrow ridge, not quite circular, rises 11,000 feet above the bottom—the height of Ætna, after which Hevel named it. It is very brilliant in Full, sometimes, according to Beer and Mädler, resembling a string of pearls, of which they once counted more than 50. A mass of ridges leans upon the wall, partly concentric, partly radiating; the latter are compared to lava, and the whole object is beautifully figured, by Herschel II. in his "Outlines of Astronomy," but he has omitted its name. It lies on the terminator a day or two after I. Quarter. Vertical illumination brings out a singular cloud composed of white streaks related to it as a centre.

CRATER-RANGES NEAR COPERNICUS. Between **110** and **112** lies one of the most curious districts in the Moon. Here, in strange contrast with the undisturbed Sinus Æstuum, Beer and Mädler have shewn 61 minute craters, believe that more than twice that number might be seen, and question whether any level ground is left. The greater part are arranged in rows; one is very evident, where they stand so close that but for their partitions it would be a canal, as indeed its N. end seems to be. The Dorpat telescope shewed Mädler, after his removal to that observatory, that great part of the Hyginus canal consists of a chain of confluent openings, and thus the idea is strengthened that these forms may have had a common origin. This region was missed by Schröter, discovered by Gruithuisen, who estimated its minute craters at 400 or 500 feet in diameter; Beer and Mädler give them about six times that size. They are easily seen, but the suitable illumination is fugitive; they should be looked for while the Sun is rising on E. side of the wall of **112**.

G

TOBIAS MAYER (**117**), 9,700 feet deep towards W., has a
fine specimen of subsequent eruption by its side.

MILICHIUS (**118**). Remarkably luminous in Full.

MARE IMBRIUM (**o**). The largest of the circular plains, ill-
defined E., five times as large as M. Crisium : one half dark
and flat, the rest very irregular.

ARCHIMEDES (**120**). One of the most regular walled plains,
60 miles in diameter, and depressed only about 650 feet; the
wall, averaging 4,200 feet above the interior, carries several
towers, the highest nearly 7,400 feet. The plain appears smooth
as a mirror, but is divided into 7 nearly parallel stripes of
unequal brightness. Gruithuisen alone has detected a minute
crater in it. The outer slope of the wall is very complex;
a magnificent object against the rising or setting Sun.

LA HIRE (**123**), a small solitary mountain, inserted in the
map of Beer and Mädler, and in the index of their "Mond,"
but strangely omitted in the text. There is nothing striking
in its usual appearance, but it was twice seen by Schröter,
under very different illumination, so brilliant as to glitter with
rays like a star : he noticed also changes in its form. Gruithuisen never saw its radiant aspect, and thought its shape
entirely altered, and its size reduced, since Schröter's time. I
once found it on the terminator, 2^d 7^h after I. Quarter, the
brightest object in sight, and radiating as described by Schröter.
On another occasion I noticed a similar hill on the other side of
122, about one-third of the distance from **122** to **121**, glittering on the terminator like a star with rays. Gerling found
the same phænomenon in a peaked hill on S. side of
Alps, 1844. I have also seen **131**, and a hill S. of it, very
brilliant in the like circumstances. Future observation may
decide whether this depends upon a peculiar angle of reflection

and vision, or on some changes in a lunar atmosphere. **123** is, according to Schröter, about 4,900 feet high.

EULER (**125**), 19 miles across and nearly 6,000 feet deep, is conspicuous for its bright streaks.

HELICON (**129**) and its neighbour Helicon A, form a curious twin crater, each 13 miles in diameter, and of vast depth. Schröter has given the latter of them 13,600 feet; its complete disappearance in Full, while **129** remains conspicuous, is a striking instance of this peculiarity, and probably accounts for the entire omission of one of them in the maps of Hevel and Riccioli.

PICO (**131**) is a very fine specimen of an insulated pyramid, rising from a narrow base to 9,600 feet according to Schröter, reduced by Beer and Mädler to 7,000 feet: on either estimate a most magnificent sight from the surrounding plain.

PLATO (**132**). A slight magnifier shews this "steel-grey" spot, lying as a lagoon at the edge of M. Imbrium, and foreshortened into an oval, whose proportions vary from (3 to 5) to (4 to 5) in the extremes of N. or S. libration, while its distance from the N. limb is only half in the one case of its amount in the other. The rampart, not a very connected one, averages 3,800 feet E., something less W., where three towers crown it, the loftiest nearly 7,300 feet, another somewhat higher bulwark surmounting the E. side. The very level interior, about 60 miles broad, is crossed according to Beer and Mädler by four lighter streaks from N. to S., and contains three little specks (one central), perhaps craters, which may be classed among the minutest objects. Gruithuisen detected seven of these. I have seen only the central one, which, once laid hold of, is not difficult under a high light. But my readers shall judge for themselves whether the local shading of the interior

is now what it was twenty-four years ago, and for this purpose
a copy from the map of Beer and Mädler (where four little
specks are shewn) is here given. *This should be carefully*

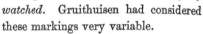

watched. Gruithuisen had considered
these markings very variable.

HARPALUS (**133**) has a wall rising
2,800 feet above the plain, nearly
16,000 feet above the chasm below.

SINUS IRIDUM (**ꝓ**), styled by Beer and
Mädler " perhaps the most magnificent of all lunar landscapes,"
is a dark semicircular bay, level almost as water, and encompassed by abrupt and colossal cliffs, the promontories being
upwards of 140 miles apart. The ridge is crowded full of
minute, usually round summits, with numberless other objects: the height is not conveniently measurable, but a point
W. of SHARP (**139**) reaches about 15,000 feet. Between it
and **132** is a strong NW. parallelism. ꝓ is in noble projection 2ᵈ or 3ᵈ after I. Quarter.

KEPLER (**144**), nearly 22 miles in diameter, sinks 10,000
feet in the deepest part beneath a very low bright wall. It is
the centre of a great ray-system, connected with that of 112.

MARIUS (**147**). More than 100 grey hillocks, the loftiest
little exceeding 1,000 feet, lie in a narrow space here.

ARISTARCHUS (**148**). The most brilliant crater in the Moon;
quite dazzling in a large telescope, the steep central hill being
the most vivid part. The ring is 28 miles across, on W. rising
7,500 feet above its inner, 2,650 feet above its outer foot: on
E. it becomes a plateau which connects it with a smaller and
steeper crater HERODOTUS (**149**). A curious serpentine valley
will be noticed here, nowhere exceeding 2½ miles in width,
not unlike a dry river-bed. The reflective power of **148** is

extraordinary, rendering it visible even to the naked eye on the bright side, and with the telescope on the dark: one of Herschel I.'s very few errors was his taking it for a volcano in eruption. Schröter pointed out this, but the mistake has been since repeated by others. There are, however, as already mentioned, variations in its light, noticed by Schröter, for which Smyth, who has seen it of every size, from a 6 to a 10 mag. star, says it is difficult to account. The bright streaks around it are no more of the nature of lava than they are elsewhere. About 45 miles WNW. begins a little group of mountains, the highest nearly 6,300 feet; these appearing on the terminator about 3ᵈ after I. Quarter are the harbingers of the great spot, conspicuous a day later.

HEVEL (154), a walled plain 70 miles across, contains a straight ridge, and an extensive convexity of surface.

ANAXAGORAS (168), 31 miles in diameter, is a very white object, including a great mountain not in the middle. It is a streak-centre.

Between TIMÆUS (170) and FONTENELLE (171) is a good deal of parallelism, and adjoining 171 on W. is one of the most curious instances of it: a nearly square enclosure foreshortened into a lozenge, whose wall-like boundaries, according to Beer and Mädler, " throw the observer into the highest astonishment." They are very unequal in height, and one is only a light streak, yet they are so regular that is scarcely possible to imagine them natural, till we find that they are 64 miles long, 250 to 3,200 feet high, and 1 mile or more thick. There are parallel ridges in the interior, and in one place the form of a perfect cross: unfortunately it lies in such a position that years, as Beer and Mädler observe, may pass without a good view of it.

E. of HORREBOW (**175**) are three curious *dark* mountains, about 5,900 feet high.

PYTHAGORAS (**176**) is the deepest walled plain in the Quadrant, nearly 17,000 feet on SE. side.

THIRD, OR SOUTH-EAST QUADRANT.

THIS includes the metropolitan crater of the Moon, TYCHO (**180**), a most perfect specimen of the lunar volcano, visible even to the naked eye in Full, and roughly figured by Galileo in the earliest telescopic representations. Its diameter is 54 miles, its depth 16,600 feet or nearly 3 miles, so that the summit of our Mont Blanc would drop beneath the ring: the height of its fine central hill 5,000 feet, ranging nearly with the terraces, from three to five rows of which border the inner slope of the narrow wall. This noble object lies among so many competitors that it is not always immediately identified by an unpractised eye near the terminator, a day or two after I. Quarter; but as the Sun rises higher upon it, its preeminence becomes more and more evident. Its vicinity is thronged with hillocks and small craters, so that for a long distance not the smallest level spot can be found; further off, the craters increase, till the whole surface resembles a colossal honey-comb. At **180** commences the largest of the systems of rays, extending over fully one-fourth of the visible hemisphere, one ray passing through M. Seren. almost to the opposite limb. Everywhere the visibility of the streaks is the converse of that of the shadows, so that many great ring-mountains which are overspread by these rays, become totally imperceptible under high illumination. They have no connection whatever with the form of the surface, are not in the least elevated or depressed, and resemble nothing on the Earth. In Full the central hill and wall of **180** are brilliant, the interior dark, as well as an exterior zone, beyond which a light cloud forms the base of the streaks. Their appearance reminds me of a vast net gathered up round a circular opening.

HESIODUS (**187**) guides the eye to a canal E. of it, in the M. NUBIUM (**s**).

CICHUS (**189**), a crater opened in table-land, lies 9,000 feet beneath the plateau, 4,000 feet below the plain : its ring is perforated by a smaller crater—an object of great interest. In 1833 I perceived that it was twice as large as it had been represented three several times by Schröter. On becoming possessed of the great map, I found it there also enlarged. Schröter, though a clumsy, was a faithful draughtsman; his views have the appearance of being each independently drawn, and they are under different angles of illumination, which often vary the size of small craters; so that here is *fair evidence of volcanic action since* 1792 : the silence of Beer and Mädler, being characteristic, goes for little or nothing. Two figures are here given, of which 1 is a copy of Schröter's design, 1792, Jan. 4,—2 is taken from Beer and Mädler's map.

LONGOMONTANUS (**192**). A great ring, 90 miles across, very deep : a peak on W. wall measures nearly 15,000 feet. The region is most wild and dislocated. **192**, with HEINSIUS (**190**), a triangle of craters each highest W., as is common here,— and WILHELM I. (**191**), are admitted by Beer and Mädler to be a little misplaced and out of proportion.

CLAVIUS (**193**). One of the grandest cavities in the Moon, though ill-placed for observation, tolerably circular, and more than 142 miles broad ; it is encompassed by a wall, damaged by successive explosions, but still portentously high and steep, attaining 17,000 feet in one of its W. peaks, and covering the gulf with night amid surrounding day. "In a good telescope," we are told by Beer and Mädler, "the prospect of a sunrise upon the surface of Clavius is indescribably magnificent." The shaded side, they justly remark, cannot be in perfect darkness when exposed to the reflection of such an enormous mass of enlightened cliffs. The bottom of the small included crater lying furthest W. falls 23,000 feet below the peak already mentioned—a space greater than the height of Chimborazo. Well may these observers express their astonishment that of this gigantic bulwark not a trace can be discovered in the Full Moon! About 1ᵈ after I. Quarter, Schröter remarks that its shadow blunts the S. horn to the naked eye.

MAGINUS (**195**). The ruin of a vast complex ring, with an interior depressed 14,000 feet.   Some minute hillocks and a little hollow in its centre form a good test. The Sun rises grandly upon it a little after I. Quarter, but "the Full Moon knows no Maginus!" S. of its W. side lies a hollow, the only dark spot under high lights in all this wild region.

SAUSSURE (**196**) interrupts one of the streaks of **180**—a solitary instance in so large a ring. W. of it the surface swarms with little craters, some hardly 0·5″ in diameter.

NASIREDDIN (**198**) and Nasireddin a, just N., are two fine craters; the latter reaches 11,000 feet. The former I have often wondered at, about I. Quarter, as the culminating point of immense explosive energy.

The GREAT CRATER-RANGE of the 1st Meridian comprises five vast walled plains in almost uninterrupted succession, which, like two other series near each limb, seem to have been formed upon a long cleft running N. and S.—similar volcanic phænomena occur on the Earth, for example, the Andes, and the Puys of Auvergne. The five lunar circles greatly resemble each other in character. We begin at the S. end.

1. WALTER (**200**) has lofty peaks on its rampart.

2. PURBACH (**202**) is about 7,500 feet deep. NE. of it, out of the main line, lies

THEBIT (**203**), 32 miles broad, 9,800 feet below one part of its ring. The wall has been pierced by a smaller and deeper crater, against the ring of which a yet lesser mine has been sprung. E. of it we find the

STRAIGHT WALL, a most curious formation, regular enough for a work of art, but more than 60 miles long, and of a very uniform height of 1,000 feet. To me it appears brownish. It begins at a small crater, and ends at a branching mountain, giving it, as Beer and Mädler say, the appearance of a staff tipped with a stag's horn. It may be well seen from 1^d to 2^d after I. Quarter. We now return to the centre of our series,

3. ARZACHEL (**204**), 65 miles across. On W. side of ring is a peak of 13,600 feet. Its neighbour

ALPETRAGIUS (**205**) is so deep—on W. 12,000 feet—as to be only five or six days free from shadow. E. of it is a very brilliant little crater, Alpetragius B.

4. ALPHONSUS (**207**), 83 miles in diameter, has a steep central peak of 3,900 feet, about the height of Vesuvius; under a high light two bright specks, and several defined blackish patches vary the surface, in those places perfectly level.

5. PTOLEMÆUS (**208**) is the last and largest of the chain : its breadth (over-stated by Schröter) is 115 miles—a magnificent lake, whose surface at sunrise or sunset is occasionally seen all roughened with ridges like waves, not 100 feet high. Pastorff has sketched it crossed by three spires of shade.

MŒSTING (**211**) has a vast depth of 7,500 feet beneath a wall only 1,600 feet high on the outside. A minute crater S., Mösting A, is very luminous. HERSCHEL (**212**), 9,500 feet deep, forms with **211** and **94** a triangle marking the Moon's centre.

BULLIALDUS (**213**), a grand crater, 38 miles across, 9,000 feet deep, with a quadruple central hill of 3,200 feet, is connected by a ravine with a little crater, and is the centre of a remarkable group.

EUCLIDES (**221**) is the best specimen of an infrequent variety, the "light-surrounded" crater. There are nine of these, all deep, regular, not large, bright, and encompassed immediately by a luminous cloud not resembling the white streaks in character. Four of them lie near

LANDSBERG (**222**), whose ring of 28 miles, with a greatest height of 9,700 feet, must command a boundless prospect over the dead level W. of it: it rises very gradually without, but is steep and terraced within.

FLAMSTEED (**223**) is a crater combined with a circle of banks averaging only 320 feet in height, which appears complete in Full.

MARE HUMORUM (**T**) is a small circular foreshortened plain, about 280 miles across, which may be detected by a keen unaided sight. The greater part is as distinctly green as M. Seren., with, in most parts, a narrow grey border.

CAMPANUS (**226**) has a remarkably dark centre; E. of it lie four little canals, three of which I have seen readily, 2ᵈ 3ʰ after I. Quarter.

VITELLO (**229**) is unparalleled for having, within a second concentric rampart, a central peak overlooking the whole ring.

GASSENDI (**232**), a conspicuous walled plain, 55 miles across, lies considerably higher than the M. Humorum. Its loftiest point is 9,600 feet above the interior, which contains many minute objects: one little hill shines brilliantly in Full. Schröter noticed several changes here. Between it and MERSENIUS (**231**) are two canals, seen plainly by me 4ᵈ 3ʰ after I. Quarter. Two little craters, Mersenius B (NE.) and c (NW.) are very luminous.

HAINZEL (**237**) is of great depth and steepness: the ring of CAPUANUS (**238**) is peculiarly irregular in height.

SCHICKARD (**239**), an enormous plain, exhibiting a fine scene between 4 and 5ᵈ after I. Quarter, is encircled by a complex wall, 460 miles round. The interior is nearly level, but its colour strongly varied; a curious contrast to the monotony of its neighbour, PHOCYLIDES (**242**).

WARGENTIN (**243**) is a singular formation; a circular elevated plain 54 miles across, excepting for a very slight rampart, resembling a large thin cheese.

BAILLY (**245**), like a small sea, lies near the limb; its wall, probably almost as high as that of **193**, seems to run back into the

DŒRFEL MOUNTAINS (**246**), a region of enormous elevation, whose summits are occasionally visible in grand profile. Beer and Mädler admit that Schröter did not over-estimate them at 25,000 or 26,000 feet.

Great craters lie between **242** and S. pole. KIRCHER (**252**)

is 18,000 feet deep, among numberless small explosions. CASATUS (**254**) is still deeper; a dome upon its mighty wall rises to about 22,300 feet. KLAPROTH (**255**) is a wonder of flatness in these regions.

NEWTON* (**256**), an irregular crater, 142 miles long, and about half as broad, is the deepest hitherto observed : the height of its loftiest tower is probably about 23,900 feet.

A colossal mountain range breaks up the limb here, fully rivalling **246**, and far exceeding anything within the disc: it extends along a considerable arc of the 3rd and 4th Quadrants. In the crescent its summits frequently prolong the horn; in Full, with suitable libration, they come out in striking projection on the sky,† while even their apparent bases may lie far above any lunar "sea." Their outline is rather rounded : some long ridges may be the profile of great rings. I find a clear description and rough measure of them by Cassini, in 1724. Schröter, who did not know of this, observed them more correctly, and named them the LEIBNITZ MOUNTAINS (**259**). The E. extremity ranges with the bright streak running S. from **180**. By some very unwonted inadvertency, Beer and Mädler have interchanged the names of **246** and **259**.‡ I have preferred retaining them as given by

---

* Beer and Mädler transferred this name from a spot so called by Schröter in the M. Imbrium close to **132**, which they thought unworthy of the designation.

† We never see a *real* Full Moon ; that is, she is never exactly opposite to the Sun except when centrally eclipsed ; hence there is always a small deficiency at one or the other pole ; and consequently not only libration but latitude must be favourable, to shew these mountains well, and *on a circular limb.*

‡ Yet they complain without cause that Schröter is not explicit here. They must have been but superficially acquainted with his work.

the earlier discoverer. These and other ranges sometimes roughen the limb during a solar eclipse, but Schröter found that their peaks then appear much sharper from the irradiation of the Sun.

BLANCANUS (**260**), a noble crater, 51 miles in diameter, with a terraced and turreted ring, domineers "like Ætna over Sicily," bearing a peak of 18,000 feet above the interior.

SCHEINER (**261**), larger and steeper, includes 10 craters and a partition.

MORETUS (**262**), a very fine object, is 78 miles broad, but of unequal height, 15,000 feet on W. side, widely terraced within, with a bright central hill, the loftiest yet measured,— 6,800 feet, by which it is easily identified. Its sunrise or sunset are very fine : it may be seen in noble relief about 1^d after I. Quarter.

GRUEMBERGER (**265**) has a central crater, whose bottom is probably more than 20,000 feet below a great peak in the wall.

ZUPUS (**268**). A valley, very dark in Full. SIRSALIS (**270**), a double ring.

GRIMALDI (**272**). The S. link of a chain of great craters lying in the meridian. This grand spot, 147 miles long by 129 wide, has a darker interior than any portion of the Moon of equal size ; it has sometimes been detected even without a telescope. Gruithuisen perceived here a regular formation somewhat resembling the letter H much inclined.

RICCIOLI (**273**) is in part as dark ; its ring is grand before the rising or setting Sun.

The CORDILLERAS (**274**), and D'ALEMBERT MOUNTAINS (**275**), a series of great ranges, nearly 20,000 feet in general height,— much higher, according to Schröter, in parts—rise along the

E. limb. They extend far S., and this extremity is called the ROOK MOUNTAINS (**276**). Two valleys were discovered by Schröter in profile on this limb, of enormous depth, rivalling the height of the mountains.

BYRGIUS (**279**) has on its ring a small crater Byrgius A remarkably brilliant, and a centre of streaks. NE. of it stands a great mountain, of probably at least 13,000 feet— as high as our Jungfrau.

FOURTH, OR SOUTH-WEST QUADRANT.

HIPPARCHUS (**288**), 92 miles across, contains all kinds of formations. Schröter has seen the shadow in its interior bordered on SW. by a fringe of fine lines, indicating a row of sharp and regular pinnacles on the wall. Three small craters, Hipparchus E, G, and C, (reckoned towards S.), in and near W. wall, are very bright in Full.

ALBATEGNIUS (**289**). A walled plain, 64 miles wide, very level. Its rampart, 14 to 18 miles broad, has been all torn by explosions. Beer and Mädler counted at least 33 of their results: a peak in NE. attains 15,000 feet. Its central hill, piercing a mass of shade, forms a noble spectacle, which I have seen 10^h before I. Quarter.

Craters with irregular walls and very large central mountains bend round from **289** through AIRY (**291**), to LA CAILLE (**292**). This last, with a wall of 9,700 feet, is quite level.

WERNER (**295**) is one of the loftiest rings; its narrow ridge ranging 13,000 feet (the height of the Alpine Eiger) above the depth, and rising on the E. to 16,500 feet, nearly 1,000 feet loftier than Mont Blanc. A spot on the SE. side of it is very bright—another, at the foot of the NE. wall, is as brilliant as **148**, and more luminous than any other part of

the Moon ; but this " star-like flashing point," being only between four and five miles square, is said to be easily overlooked
with low powers. I have several times readily seen it with two
achromatics of $3\frac{7}{10}$ inches aperture, and powers 75, 80, and
144, but *never of the specified brilliancy :* it may deserve
attention.

ALIACENSIS (**296**) is similar to **295**, but larger.

THEON SEN. (**297**), THEON JUN. (**298**), ALFRAGANUS (**300**), are
all luminous.

ABULFEDA (**305**) and ALMANON (**306**) have their walls
united by a row of little craters like a canal, visible about
I. Quarter.

ABENEZRA (**310**) sinks more than 14,500 feet. PONS (**313**)
has very dark spots in the ring.

ALTAI MOUNTAINS (**315**). This long range (whose name is
omitted in Beer and Mädler's map) is almost the only undisturbed high ground in this Quadrant. Its rounded summits lie
awkwardly for measurement, but seem about 13,000 feet. I
have often seen the NW. face as a bright line of·nearly 280
miles of cliffs, in the increasing Moon.

A group of three huge craters follows, forming a wild and
gigantic region, very difficult to delineate. The details occupied Beer and Mädler portions of more than fifty nights.

THEOPHILUS (**319**) is the deepest of all visible craters, if we
regard the general line of the ring, which ranges from 14,000
to 18,000 feet above the chasm, with a diameter of 64 miles.
No scene in the least approaching to it exists on the Earth.
The central peak is 5,200 feet high. CYRILLUS (**320**), equally
large and terraced, approaches to a square. A small crater
Cyrillus A, on its E. side, is very brilliant. CATHARINA (**321**)
is rather the largest of the three, and more than 16,000 feet

deep, but irregular in character.   Of this superb cluster,
319 first catches the rising Sun, and I have seen it far beyond
the terminator, and even without the telescope, 5^{d} after New:
it is a grand object when filled with night, through which its
glittering central peak comes out like a star.   A wide valley
connecting **320** and **321** is much better represented in the
beautiful old map of Tobias Mayer, and by Russell, than by
Beer and Mädler.

ISIDORUS (**323**) and CAPELLA (**324**) lie side by side, with
a peak of more than 13,000 feet between them; three clefts
cut down the ring of **324**.   Schröter here saw traces of a line
of little confluent craters.   Double craters, too numerous not
to have some special cause, abound in this region.

CENSORINUS (**325**), a minute crater, with its vicinity, is very
brilliant in Full.

MESSIER (**327**) and Messier A, a pair of small deep
craters, exhibited in the time of Beer and Mädler such a
curious similarity in size, form, depth, brightness, and even
the position of the peaks upon their rings, that it must have
been either a wonderful coincidence, or the result of some
unknown natural process.   I have as yet been unable to get
a thoroughly satisfactory view of them with adequate optical
means, but believe I am warranted in saying that this similarity no longer exists, and that we have in this place decided
evidence of *modern eruptive action*.   Two curious white
streaks, slightly divergent, extend from Messier A for a
long distance E., forming with the included shade the picture
of a comet's tail.   Gruithuisen, who imagined them to be artificial, states that they are composed of a multitude of distinct
parallel lines.   In consequence of an observation by Schröter,
who discovered this " comet," Beer and Mädler fortunately

examined this spot, so peculiarly calculated to exhibit any variation, *more than* 300 *times*, between 1829 and 1837, without noticing any change. 1855, Nov. 14, I perceived with my 5½ feet achromatic, that the E. crater appeared the larger of the two. 1856, March 11, I found the W. crater not only the lesser, but "*lengthened obviously in an E. and W. direction.*" I have since noted the same dissimilarity with a smaller as well as two much larger instruments, though I have never yet had a really clear view: the last time, 1859, Feb. 8, with a 5½ inch object-glass, by the celebrated Alvan Clark, I thought W. crater much shallower than E. The figure is taken from a rough sketch, 1857, Feb. 28: the shadow is probably too broad in W. crater, which is that to the left in the inverted diagram.

GUTTEMBERG (**330**), like many of its neighbours, is pear-shaped. E. and S. of it lie the

PYRENEES* (**331**), two mountain masses, the N. portion 12,000 feet high.

BORDA (**337**) has a lofty range on its W. side; a peak (Borda *a*) lying in the direction of the centre of Petavius rises at one spring 11,000 feet; similar in abruptness to the Alpine Wetterhorn, or the two great Pics du Midi in the terrestrial Pyrenees.

We come now to the last of the three great rows of craters lying in the meridian, consisting of the four following objects, which must be looked for in the crescent 3ᵈ or 4ᵈ old, or at a shorter interval after Full.

* This name is omitted in the " Mappa Selenographica "

1. LANGRENUS (**338**) is a superb object under oblique sunshine, with its multiple ring 9,600 feet high, and a brilliant central hill.

2. VENDELINUS (**339**) is unequal to **338**, though on the Earth it would be esteemed a wonder of magnificence; it contains a very dark speck in Full.

3. PETAVIUS (**340**) is one of the finest spots in the Moon: its grand double rampart, on E. side nearly 11,000 feet high, its terraces, and convex interior with a central hill and canal, compose a magnificent landscape in the lunar morning or evening, entirely vanishing beneath a Sun risen but half way to the meridian.

4. FURNERIUS (**345**) has a central crater: the deep little crater Furnerius A at N. end of the wall is, in Full, a brilliant point on a white streak, and with another streak E. of STEVINUS (**344**) identifies this neighbourhood. On the limb are some huge rings. WILHELM HUMBOLDT (**352**) shews peaks of 16,000 feet in profile.

STŒFLER (**354**) rising in one part to 12,000 feet, has a very level interior, crossed by two of the rays from **180**. It is well seen about I. Quarter.

MAUROLYCUS (**358**). A noble walled plain, sure to be found in grand relief about I. Quarter, or 2^d before III. Quarter. Its complex rampart is heaved up on E. side to 13,800 feet—on W., though not according to Beer and Mädler quite so lofty, I have often seen it more conspicuous in the increasing Moon. The same observers describe the plain as crossed by 12 diverging bright lines in Full.

The ring of BACON (**360**) has a peak on E. side nearly as high as **358**, and on the top of its SE. side a row of five minute craters.

LINDENAU (**370**), a perfect ring, rises on the E. by four stages.

PICCOLOMINI (**371**), a noble circle 57 miles in diameter, has a central hill and complex wall, bearing on E. a tower about the height of our Mont Blanc.

FRACASTORIUS (**372**), an imperfect ring, forms a bay of the MARE NECTARIS (**v**): a little crater on E. headland is brilliant in Full.

From W. side of REICHENBACH (**375**) towards RHEITA (**376**) extends one of the gigantic valleys peculiar to this region. **376** has a ring reaching 14,000 feet. Close under it, on the side next METIUS (**384**) is another valley 186 miles long, and from 14 to 23 miles wide, bordered by steep walls; in one spot more than 9,000 feet deep. Along W. side of FRAUENHOFER (**377**) is another valley 7 miles wide, which may be traced through **345**, though of very unequal depth, for 212 miles.

STEINHEIL (**385**) is probably the deepest of the double rings, in one place sinking to 12,000 feet.

VLACQ (**388**) the largest of an enormous group, is 57 miles in diameter.

A little crater E. of NICOLAI (**393**) called Nicolai a, is very bright in Full. Minute insulated craters abound here at the rate of 65,000 for the visible hemisphere.

ZACH (**396**), 13,000 feet deep, is magnificently terraced.

CURTIUS (**404**) has a very complex and steep wall, probably surpassing the elevation of our Chimborazo.

INDEX TO THE MAP OF THE MOON.

GREY PLAINS, USUALLY CALLED SEAS.

A.	Mare	Crisium.	**M.**	Sinus	Medii.
B.	—	Humboldtianum.	**N.**	—	Æstuum.
C.	—	Frigoris.	**O.**	Mare	Imbrium.
D.	Lacus	Mortis.	**P.**	Sinus	Iridum.
E.	—	Somniorum.	**Q.**	Oceanus	Procellarum.
F.	Palus	Somnii.	**R.**	Sinus	Roris.
G.	Mare	Tranquillitatis.	**S.**	Mare	Nubium.
H.	—	Serenitatis.	**T.**	—	Humorum.
I.	Palus	Nebularum.	**V.**	—	Nectaris.
K.	—	Putredinis.	**X.**	—	Fœcunditatis.
L.	Mare	Vaporum.	**Z.**	—	Australe.

CRATERS, MOUNTAINS, AND OTHER OBJECTS.

1.	Promontorium Agarum.	**32.**	Franklin.
2.	Alhazen.	**33.**	Berzelius.
3.	Eimmart.	**34.**	Hooke.
4.	Picard.	**35.**	Strabo.
5.	Condorcet.	**36.**	Thales.
6.	Azout.	**37.**	Gärtner.
7.	Firmicus.	**38.**	Democritus.
8.	Apollonius.	**39.**	Arnold.
9.	Neper.	**40.**	Christian Mayer.
10.	Schubert.	**41.**	Meton.
11.	Hansen.	**42.**	Euctemon.
12.	Cleomedes.	**43.**	Scoresby.
13.	Tralles.	**44.**	Gioja.
14.	Oriani.	**45.**	Barrow.
15.	Plutarchus.	**46.**	Archytas.
16.	Seneca.	**47.**	Plana.
17.	Hahn.	**48.**	Mason.
18.	Berosus.	**49.**	Baily.
19.	Burckhardt.	**50.**	Burg.
20.	Geminus.	**51.**	Mt. Taurus.
21.	Bernouilli.	**52.**	Römer.
22.	Gauss.	**53.**	Le Monnier.
23.	Messala.	**54.**	Posidonius.
24.	Schumacher.	**55.**	Littrow.
25.	Struve.	**56.**	Maraldi.
26.	Mercurius.	**57.**	Vitruvius.
27.	Endymion.	**58.**	[Mt. Argæus].
28.	Atlas.	**59.**	Macrobius.
29.	Hercules.	**60.**	Proclus.
30.	Oersted.	**61.**	Plinius.
31.	Cephesu.	**62.**	Ross.

63. Arago.
64. Ritter.
65. Sabine.
66. Jansen.
67. Maskelyne.
68. Mt. Hæmus.
69. Promontorium Ache-
rusia.
70. Menelaus.
71. Sulpicius Gallus.
72. Taquet.
73. Bessel.
74. Linné.
75. Mt. Caucasus.
76. Calippus.
77. Eudoxus.
78. Aristoteles.
79. Egede.
80. Alps.
81. Cassini.
82. Theætetus.
83. Aristillus.
84. Autolycus.
85. Apennines.
86. Aratus.
87. Mt. Hadley.
88. Conon.
89. Mt. Bradley.
90. Mt. Huygens.
91. Marco Polo.
92. Mt. Wolf.
93. Hyginus.
94. Triesnecker.
95. Manilius.
96. Julius Cæsar.
97. Sosigenes.
98. Boscovich.
99. Dionysius.
100. Ariadæus.
101. Silberschlag.
102. Agrippa.
103. Godin.
104. Rhæticus.
105. Sömmering.
106. Schröter.
107. Bode.
108. Pallas.
109. Ukert.
110. Eratosthenes.
111. Stadius.

112. Copernicus.
113. Gambart.
114. Reinhold.
115. Mt. Carpathus.
116. Gay Lussac.
117. Tobias Mayer.
118. Milichius.
119. Hortensius.
120. Archimedes.
121. Timocharis.
122. Lambert.
123. La Hire.
124. Pytheas.
125. Euler.
126. Diophantus.
127. Delisle.
128. Carlini.
129. Helicon.
130. Kirch.
131. Pico.
132. Plato.
133. Harpalus.
134. Laplace.
135. Heraclides.
136. Maupertuis.
137. Condamine.
138. Bianchini.
139. Sharp.
140. Mairan.
141. Louville.
142. Bouguer.
143. Encke.
144. Kepler.
145. Bessarion.
146. Reiner.
147. Marius.
148. Aristarchus.
149. Herodotus.
150. Wollaston.
151. Lichtenberg.
152. Harding.
153. Lohrmann.
154. Hevel.
155. Cavalerius.
156. Galileo.
157. Cardanus.
158. Krafft.
159. Olbers.
160. Vasco de Gama.
161. Hercynian Mts.

162.	Seleucus.		**211.**	Mösting.
163.	Briggs.		**212.**	Herschel.
164.	Ulugh Beigh.		**213.**	Bullialdus.
165.	Lavoisier.		**214.**	Kies.
166.	Gérard.		**215.**	Gueriké.
167.	Repsold.		**216.**	Lubiniezky.
168.	Anaxagoras.		**217.**	Parry.
169.	Epigenes.		**218.**	Bonpland.
170.	Timæus.		**219.**	Fra Mauro.
171.	Fontenelle.		**220.**	Riphæan Mts.
172.	Philolaus.		**221.**	Euclides.
173.	Anaximenes.		**222.**	Landsberg.
174.	Anaximander.		**223.**	Flamsteed.
175.	Horrebow.		**224.**	Letronne.
176.	Pythagoras.		**225.**	Hippalus.
177.	Œnopides.		**226.**	Campanus.
178.	Xenophanes.		**227.**	Mercator.
179.	Cleostratus.		**228.**	Ramsden.
180.	Tycho.		**229.**	Vitello.
181.	Pictet.		**230.**	Doppelmayer.
182.	Street.		**231.**	Mersenius.
183.	Sasserides.		**232.**	Gassendi.
184.	Hell.		**233.**	Agatharchides.
185.	Gauricus.		**234.**	Schiller.
186.	Pitatus.		**235.**	Bayer.
187.	Hesiodus.		**236.**	Rost.
188.	Wurzelbauer.		**237.**	Hainzel.
189.	Cichus.		**238.**	Capuanus.
190.	Heinsius.		**239.**	Shickard.
191.	Wilhelm I.		**240.**	Drebbel.
192.	Longomontanus.		**241.**	Lehmann.
193.	Clavius.		**242.**	Phocylides.
194.	Deluc.		**243.**	Wargentin.
195.	Maginus.		**244.**	Inghirami.
196.	Saussure.		**245.**	Bailly.
197.	Orontius.		**246.**	Dörfel Mts.
198.	Nasireddin.		**247.**	Hausen.
199.	Lexell.		**248.**	Segner.
200.	Walter.		**249.**	Weigel.
201.	Regiomontanus.		**250.**	Zuchius.
202.	Purbach.		**251.**	Bettinus.
203.	Thebit.		**252.**	Kircher.
204.	Arzachel.		**253.**	Wilson.
205.	Alpetragius.		**254.**	Casatus.
206.	Promontorium Æna-		**255.**	Klaproth.
	rium.		**256.**	Newton.
207.	Alphonsus.		**257.**	Cabeus.
208.	Ptolemæus.		**258.**	Malapert.
209.	Davy.		**259.**	Leibnitz Mts.
210.	Lalande.		**260.**	Blancanus.

357. Clairaut.	**381.** Pontécoulant.
358. Maurolycus.	**382.** Hanno.
359. Barocius.	**383.** Fabricius.
360. Bacon.	**384.** Metius.
361. Buch.	**385.** Steinheil.
362. Büsching.	**386.** Pitiscus.
363. Gemma Frisius.	**387.** Hommel.
364. Poisson.	**388.** Vlacq.
365. Nonius.	**389.** Rosenberger.
366. Fernelius.	**390.** Nearchus.
367. Riccius.	**391.** Hagecius.
368. Rabbi Levi.	**392.** Biela.
369. Zagut.	**393.** Nicolai.
370. Lindenau.	**394.** Lilius.
371. Piccolomini.	**395.** Jacobi.
372. Fracastorius.	**396.** Zach.
373. Neander.	**397.** Schomberger.
374. Stiborius.	**398.** Boguslawsky.
375. Reichenbach.	**399.** Boussingault.
376. Rheita.	**400.** Mutus.
377. Frauenhofer.	**401.** Manzinus.
378. Vega.	**402.** Pentland.
379. Marinus.	**403.** Simpelius.
380. Oken.	**404.** Curtius.

MARS.

THIS planet, which is only about twice the size of the Moon, and not much more than half as large as our own globe, is yet peculiarly interesting to us, as presenting the most intelligible features of any object within our reach. In overtaking him about once in two years we find, as he turns to us his round sunny face, that his supposed malignant aspect is changed into that of a miniature Earth, which we might, without much extravagance, imagine to be habitable by man. Not every *opposition*, however, as it is called, admits of an equally near prospect. The orbits of both the Earth and Mars are elliptical, and not fixed with respect to each other, and no two following oppositions happen in the same part of either orbit, so that the most favourable possible juncture, when the Earth is furthest from the Sun and Mars nearest, occurs ordinarily but once in 15 years, when the diameter of Mars, only 13″ in reversed circumstances, expands to 23·5″. Every opposition, however, should set the telescope to work; and we will proceed to describe what we may expect to see.

1. The *Phases*. These are not remarkable : in opposition, a full moon rising through ruddy haze, and, with sufficient power, larger than our Moon to the naked eye : in other situations a dull gibbosity, never sinking to quadrature. Mädler stated at one time that this phasis is always narrower than it should be by calculation; but in a subsequent publication the remark is not repeated. Pastorff thought he saw a phosphorescence on the dark part : but he was not a very good observer.

2. The *Dark Spots*. The disc, when well seen, is usually mapped out in a way which gives at once the impression of land and water : * the bright part is orange,—according to Secchi, dotted with brown and greenish points; Beer and Mädler think it much less red than to the naked eye : the darker spaces are of a dull grey-green,† or according to Secchi, bluish, possessing the aspect of a fluid absorbent of the solar rays. If so, the proportion of land to water on the Earth is reversed on Mars ; on the Earth every continent is an island ; on Mars all seas are lakes ; (and those, according to Jacob and Secchi, like our own continents, chiefly confined

* Secchi's illustration is strikingly expressed : "è tutto variegato come una carta geografica."

† Herschel II. refers this colour to contrast. Jacob does not see it. The accurate Humboldt has puzzled himself about the colours on Mars. (Cosmos, IV. 503.)

to one side of the globe ;) so that the habitable area may possibly be much more alike than the size of the planets. The
dark spots were early seen, and a long series of drawings is
extant from Hooke, Cassini, and Campani, in 1666,* to Jacob,
Secchi, and De La Rue in the present day, with some general
correspondence,† but wide difference of detail ; this may be
due to differences in telescopes, eyes, and climates, and still
more to the great real change owing to the inclination of the
axis,‡ shewing us sometimes more of N., sometimes of S.
hemisphere : still the explanation seems inadequate, and we
might at first doubt whether our land and sea were not vanishing into nothing more than vapours. The older observers
thought the spots variable: Herschel I. perhaps took the lead
in supposing them to be permanent, an idea which Kunowsky,
as recently as 1822, fancied was due to himself. Schröter's
work on Mars, the "Areographische Fragmente," which was
to have contained 224 figures, was unfortunately left in MS.
at his death in 1816, but he has stated that he and Olbers
found them vary rapidly. Beer and Mädler took up the subject with great spirit at the peculiarly favourable opposition in
1830, recovered some of Kunowsky's spots, and from their

* Humboldt, following Delambre, says that Cassini does not seem to
have discovered the rotation of the spots till after 1670. He must have
overlooked the figures in Phil. Trans. No. 14. Kaiser has shewn, from
the MS. journal of Huygens, that the latter discovered the rotation in
1659.

† Herschel I.'s general figure in Phil. Trans. 1784, if *reversed*, will be
not unlike Beer and Mädler's polar projections.

‡ 30° 18', according to Beer and Mädler, and Herschel II. But is
not this a mistake for the complementary angle ? Hind gives, after
Herschel I., 61° 18' for inclination of axis to orbit of Mars, 59° 42' for
do. to Earth's ecliptic, obliquity on Mars, 28° 42' : but 90° — 59° 42 =
30° 18'.

further observations in 1832, 1834, and 1837, though the same hemisphere was not always equally visible, inferred their permanence. Mädler, however, seems a little shaken in 1839, and retrogrades still further at Dorpat in 1841; nor do the drawings of later observers exhibit the same forms, though they seem in general persuaded that the spots are really part of the surface. However, the distant view of the Earth might be much of this nature; its outlines at one time distinct, at another confused or distorted by clouds : besides, one affirmative—the re-appearance of a spot—proves more, where there may be hindrances, than can be disproved by many negatives. On the whole, we may presume that they are permanent; though it may be a long time before we can form a good map of Mars, nor shall we ever know the N. so well as the S. hemisphere, as it is turned towards us in the planet's aphelion;—even were its markings equally defined,

which Beer and Mädler deny. Secchi has recently promised a special memoir, which is sure to be most valuable. Two of his figs. are here inserted from the last vol. of the Astron. Nachrichten.—That to the left represents the planet, 1858,

June 3, 9^h 45^m P.M.—the other, June 14, 9^h 15^m. The letters *a a* point out the continuation of the same luminous continent. Under favourable circumstances the dark spots are not difficult objects, and I have repeatedly been able to draw them with my $5\frac{1}{2}$ feet achromatic. Their motion will be very evident, and as the rotation is completed, according to Beer and Mädler, in 24^h 37^m 24^s, they will not vary greatly from night to night at the same hour.

3. The *Polar Snows*. A circular spot on each hemisphere is so white and luminous as to have alone remained visible to Beer and Mädler when a cloud obscured the planet: and occasionally to seem, from irradiation, to project beyond the limb, as I have myself noticed. These zones were figured by Maraldi in 1704, who says they had been occasionally seen for 50 years; * in fact they could not long escape the telescope. They were thought to resemble snow before Herschel I.'s time; he gave consistency to the idea by ascertaining that they decreased during the summer, and increased during the winter of Mars, and Beer and Mädler have fully confirmed it, with the addition that the S. polar spot has a greater variety of extent, corresponding with its greater variety of climate from the excentricity of the orbit. Each pole comes alternately into sight, and both are sometimes visible on the edge at once, when the opposition of Mars concurs with his equinox. Herschel I. found they were not (or not always) opposite each other, both being sometimes in or out of the disc at the same time. Mädler, and Secchi with the admirable achromatic at Rome, of $9\frac{6}{10}$ inches aperture

---

* Beer and Mädler erroneously make him the discoverer in 1716. A figure by Huygens, in 1656, seems intended as a rude representation of them.

and 15 feet focus, bearing ordinarily a power of 1,000, found the N. zone concentric with the axis, but the S. considerably excentric. It has been suggested by Beer and Mädler, that the poles of cold, like those on the Earth, may not coincide with the poles of rotation;—still they should be diametrically opposite. These observers found in 1837 the N. pole surrounded by a conspicuous dark zone, the only well-marked spot in sight, which they thought might possibly be a marsh at the edge of the melting snow : in 1839 Mädler perceived it had decreased ; in 1841 it was no longer visible. About the opposition in 1856 I had interesting views of these zones, which did not seem exactly opposite to each other : the S. was surrounded by a very dark region, never seen by Beer and Mädler ; on the intervening limbs were occasionally luminous regions, so bright by contrast as to give an impression of *four* patches of snow, as in one of Cassini's figures in 1666 ; these were also seen by Secchi at the same time.   In 1845 Mitchell with a great achromatic in America noticed a very dark spot in the centre of the snow, which disappeared the next night ; at another time he saw some movements in a small bright spot at the edge of the snow.   On the whole we must adopt Secchi's view, who, with every advantage of sky and telescope, thinks much investigation is here required.

4. The *Atmosphere*.   Such an appendage is implied in the formation of snow, the uncertain shape of the dark spots, and their usual disappearance towards the limb : Beer and Mädler found them also better defined in the summer than the winter of Mars.   Maraldi saw and delineated a dusky belt for a considerable time in 1704; something like it also appears in the designs of Schröter, who, from movements in these belts, inferred winds as rapid as our own.   Some of Herschel I.'s figures shew

white belts, and he says that besides the permanent spots he often noticed occasional changes of partial bright belts, and once a darkish one; we should however, perhaps, have expected that clouds would always reflect a brighter light than land or water. The bright "menisci" or crescents which some observers have seen illuminating the E. and W. borders of the disc may have had an atmospheric cause, as well as the numerous patches of yellowish and bluish light upon the limb described by Mädler, at Dorpat, in 1841. The atmosphere is probably dense, but Cassini exceeded all bounds in supposing it could obscure small stars at some distance; this effect, resulting from the contraction of the pupil in a bright light, was imperceptible in the great telescopes of Herschel I., and the idea has been overthrown by the experience of South, who has seen one contact and two occultations of stars without change: in the last, his great achromatic, $11\frac{7}{8}$ inches aperture and nearly 19 feet focus, actually shewed the star neatly dichotomized in emerging. This is not surprising, for the atmosphere, if in proportion to ours, would not extend more than 0·3″ beyond the limb when nearest to the Earth.

Favourable oppositions will occur in 1860, 1862, and 1877.

———————

The MINOR PLANETS hardly come within the design of these pages, though the more conspicuous of them may sometimes be found by the Nautical Almanac without much trouble But any investigation respecting the discs, nebulous envelopes, and irregular light of a few of these extremely minute bodies belongs only to first-rate instruments.

JUPITER.

THIS magnificent planet is an excellent object, in view for months together, and for some part of every year, shining with a brilliancy which near the opposition casts shadows in a darkened room, and has been compared to that of a coach-lamp in the Earl of Rosse's reflector; it is attended, too, by a most interesting retinue. In the latter even a very moderate telescope will shew one of the most diversified scenes in the heavens. Not only from night to night, but often from hour to hour, incessant changes will catch the eye; and since the orbits of all the satellites lie nearly edgeways with respect to the Earth, the retreating and advancing motions appear to intersect one another, and these miniature planets are seen overtaking, passing, meeting, hiding, and receding from one another in most beautiful and endless mazes. We begin with the features of the globe.

1. The *Ellipticity.* No heavenly body (except at times Venus) shews a disc so readily as Jupiter; a very low power will bring out his noble face, 11 times larger than our Earth; and with higher magnifiers we shall soon perceive that it is not circular, but a little flattened N. and S.* This is very obvious, yet it was doubted at one time by Cassini, and missed by Hooke, though he drew the spots of Mars, and directed particular attention to the polar flattening of the Earth. Astronomers differ as to its amount; from many comparisons and

---

* The beginner may be reminded that the points of the celestial compass only stand straight when the object is on the meridian; but at all times they may be verified by recollecting that the motion through the field is always from E. to W.

observations Main fixes it at a little more than $\frac{1}{17}$* of the equatorial diameter.   Once noticed, it is so evident that a good eye will not tolerate the circular figures too often used to represent this planet : and the student who wishes to draw the belts had better prepare oval discs beforehand, which may be done thus. Make a rectangle 16 high, 17 wide, on any convenient scale of equal parts; find its centre by intersecting diagonals; describe a circle on this centre touching the top and bottom, and then pull out as it were the sides of the circle to touch the ends of the rectangle, altering the curves by eye and hand till a tolerable ellipsis is produced : if many figures are wanted, cut one out neatly in card, and draw the others by means of its edge.

   2. The *Phasis.*   The orbit of Jupiter is so far external to that of the Earth that there can be but little defalcation of light — ordinary books say *none* : a diminished breadth is however measurable, if not visible, in our best instruments, and a moderate telescope will show the *approach* of the gibbous form about the time of Jupiter's quadrature, in a slight shade along the limb furthest from the Sun.   This was recorded by myself as far back as 1838, May 30, but I believe I had seen it previously, as often since, with the $5\frac{1}{2}$ feet achromatic : it has recently been taken notice of by De La Rue and Secchi. I have thought it plainer in twilight than darkness; a fact in accordance with many similar observations by eminent astronomers.†

   3. The *Belts and Spots.*   The early telescopes must soon

* Secchi's grand achromatic and pure sky give $\frac{1}{16}$.

† The most perfect telescopic vision is sometimes attained under unlikely circumstances,— such as fog, twilight, and moonlight : every object is best seen under certain proportions of light and power and contrast, which are matter of experience ; and in such experience lies much of the observer's skill.

have shewed grey streaks across the disc; and it would be
found in time that, though always lying in one equatorial
direction, they were not permanent, but subject to sometimes
very gradual, sometimes very rapid changes. Of the latter
Herschel I. gives the following extreme instance. "I had
been observing Jupiter for some time with the 20 feet reflector;
I never saw him more satisfactorily; he was covered with
belts. My wife sent for me to take my tea, I left my telescope with Jupiter in the field of it covered, absolutely
covered with belts. After I had taken my tea, I returned to
my observations; but on bringing Jupiter again into the field
of the telescope, no vestige of belts could I detect. The
time I was absent from my telescope was not more than 20ᵐ."
Such cases are rare; but changes, on a vast scale when we
bear in mind the size of the planet, often may be traced
at short intervals. The equator is generally luminous, having
on each side of it a dusky streak, beyond which a series of
narrower stripes extend to either pole: but the arrangement
is very uncertain in all respects except that of constant E. and
W. direction; oblique, curved, or ragged streaks are however
sometimes seen, and the belts are frequently interrupted, as
well as varied by waved and festooned outlines and looped
appendages. They (the large belts especially) have a brownish or coppery hue: of late they have been singularly faint.
Spots, equally uncertain, are from time to time visible. They
are occasionally luminous, more frequently very dark:
one of the latter, seen by Hooke in 1664 and Cassini in
1665, was unusually permanent, having been observed, with
many interruptions, till 1715; it even seems to have re-
appeared as late as 1834,* always with the S. equatorial

* According to Hind, however, these spots were on opposite sides of

I

belt, though the belt has often been seen without it. An observation by South affords a beautiful illustration of the evanescent nature of some of these objects. " On June 3, 1839, at 13ʰ 45ᵐ (sidereal time) I saw with my large achromatic, immediately below the lowest [? edge] of the principal belt of Jupiter, a spot larger than I had seen before: it was of a dark colour, but certainly not absolutely black. I estimated it at a fourth of the planet's equatorial diameter. I shewed it to some gentlemen who were present; its enormous extent was such that on my wishing to have a portrait of it, one of the gentlemen, who was a good draftsman, kindly undertook to draw me one: whilst I, on the other hand, extremely desirous that its actual magnitude should not rest on estimation, proposed, on account of the scandalous unsteadiness of the large instrument, to measure it tricometrically (*sic*) with my 5 f. equatorial. Having obtained for my companion the necessary drawing instruments, I went to work, he preparing himself to commence his; on my looking however into the telescope of the 5 f. equatorial, at 13ʰ 45ᵐ (*sic*), I was astonished to find that the large dark spot, except at its eastern and western extremities, had become much whiter than any of the other parts of the planet, and at 14ʰ 19ᵐ these miserable scraps were the only remains of a spot which but a few minutes before, had extended over at least 22,000 miles." A very different kind of spots has recently been observed;— minute white roundish specks, about the size of satellites, on the dark S. belts. Dawes first saw them in 1849, Lassell in

the equator. The spot, or rather spots, of 1834 had been seen by Schwabe in 1828, and by myself with a fluid achromatic in 1831 and 1832. Dawes observed a very large black spot in 1843. Two were visible at the close of 1858.

1850, with his reflector, one the finest instruments in the world, 2 feet aperture, 20 feet focus. Dawes has since given several striking drawings of them in the Monthly Notices of the Astronomical Society, and they have been seen in the great achromatic of 9 inches aperture recently erected by Sir W. K. Murray in Scotland. They are evidently not permanent. Common telescopes will have no chance with them, or with the similar traces which Lassell has detected on the bright belts. All these phænomena prove an atmosphere vaporous and mutable like that of the Earth; even in our own atmosphere, when near the " dew point" or limit of saturation with moisture, surprising changes sometimes occur: a cloud-bank observed by Herschel II, 1827, April 19, was precipitated so rapidly that it crossed the whole sky from E. to W. at the rate of at least 300 miles per hour; and alterations far more sudden are conceivable where everything is on a gigantic scale. The ancient astronomers soon took this view, and Huygens speaks of clouds and winds on Jupiter; but they seem to have looked upon the dark as the cloudy belts, forgetting that to an eye placed *above them*, vapours in sunshine would appear whiter than the globe beneath. Schröter made the same mistake, and Herschel I. for a time, though subsequently he perceived the truth, that the dusky belts are the real body of the planet: this is confirmed by the invisibility of the darkest spots towards the limbs, and the fading away of the grey streaks towards their ends;—an appearance which I never could quite satisfy myself of, but which seems verified by larger telescopes,* and must indicate great density of

* Herschel I. does not mention it: his son describes and represents it: it is greatly exaggerated in the clumsy figures of Beer and Mädler, and does not appear in the exquisite design of De La Rue.

atmosphere; this cannot however be of much depth, or it would be made apparent when the satellites pass behind the disc. The equatorial direction indicates a " trade-wind," more immediately the result of rapid rotation than the currents of the Earth: several observers have noticed the usual forms of clouds, and Piazzi Smyth, in his recent charming book " Teneriffe," has described in the most graphic manner the appearance of " a windy sky " and the shapes of drifting and changing vapours, seen with the grand achromatic of $7\frac{1}{4}$ inches aperture on "the peak of Teyde," and beautifully delineated in his more scientific " Report." The great dark spots possess more stability, as though some portions of the surface cleared the sky above them for a considerable time; yet these, or rather the vapours round them, seem liable to displacement, as they do not always give the same period of rotation. This however, according to Beer and Mädler, is very nearly 9^h 55^m 30^s for day and night.* The equator of this huge globe is therefore flying 28,000 miles an hour, or between 7 and 8 miles every second! and a few minutes shew the movement of the spots, but puzzle the draughtsman. We must now proceed to

THE SATELLITES.

Are these little moons visible to the naked eye? The question has been usually negatived, as, though I (i.e. the first, or nearest) is about as far from Jupiter's surface as our Moon from the Earth, they are very minute, and over-powered by the planet's rays. Yet there have been exceptions. A tailor, named Schön, at Breslau, who died in 1837, always and indisputably perceived I and III, when sufficiently distant from Jupiter. The Marquis of Ormonde is said to have seen them in the sky of Etna: Jacob and another have made out

---

* Airy, from Cambridge observations in 1834, makes it 9ˢ less.

III at Madras: the missionary Stoddart at Oroomiah, in
Persia, states that he could detect some of them in twilight,
before the glare of the planet came out; and under the same
circumstances two are said to have been very recently perceived by Mr. Levander and others at Devizes: * once (1832,
Sept. 1), III and IV being on the same side, and far from
Jupiter, I saw them, though not separately, through the concave eye-glass which corrects my near sight. To try this
experiment, the planet had better be just hid behind some
object, though I did not find it necessary: but any small
telescope will bring them out, and a powerful instrument will
even shew them in the day-time. Like Galileo, we may not
at first perceive all four, as some are often invisible before or
behind Jupiter, or in his shadow, but we shall soon recognise
the whole train, and must attend to the following particulars.

1. Their *Identification*, a matter of some importance. III,
if well situated, is usually brightest at first sight. IV goes
much further than the others: but the best way is to consult
the daily configurations in the Nautical or Dietrichsen's Almanac, attending to the explanation of the symbols.

2. Their *Magnitudes*. Even a small instrument will shew
that their light is steadier than that of stars; this arises from
their possessing real discs, which will soon be "raised" or
drawn out, with increase of aperture and power; and beautiful miniature full moons they will be found to be; by no
means, however, of one size. Struve's last corrected measures
give I, 1·015″—II, 0·911″—III, 1·488″—IV, 1·273″.† This

* This occurred about 20ᵐ before a remarkable crimson aurora, 1859,
April 21. Compare the observation of Venus, p. 43. Herschel I. found
no perceptible effect from auroral light.

† Secchi gives 0·985″, 1·054″, 1·609″, 1·496″. The two first measures
are due to a single day; this anomaly will be again mentioned.

accounts in part for their different brightness, but a difference
of reflective power must be combined with it in the general
effect: and the individual light of each varies at different
times. From many comparisons Herschel I. and Schröter
considered that, like our Moon, they always turn the same
side to their primary, and consequently different faces—some
of which may be darkened by spots—to us: this has been
confirmed by Beer and Mädler, but is still an inadequate explanation; its results are not always uniform, and singular
anomalies occur, especially with IV. As far back as 1707
Maraldi noticed that, though usually faintest, it was sometimes brightest; in 1711 Bianchini and another once saw it
for more than 1^h so feeble that it could hardly be perceived;
1849, June 13, Lassell made the same observation with
perfect accuracy. III is more consistent, usually taking the
lead, yet Maraldi and Bond have sometimes observed the
contrary; and many years ago, when I paid some attention
to this subject, I have seen it repeatedly surpassed by IV.
Those who would study these changes must hide the planet
behind a narrow bar in the field, made by temporarily removing the field-glass, and placing a thick wire, or strip of metal,
or wood, or card, against the diaphragm which forms the
edge of the field: it is also advantageous to throw the satellites a little out of focus, as thus we *eliminate*, as mathemati-
cians say, or get rid of, the impression of size, which might
mislead the eye. Herschel I. used a convenient mode
of expressing differences of light by stops of different value;
thus III : I II, IV would signify that III was very much
brighter than I and II, and these again a very little brighter
than IV. The distances from Jupiter must be estimated in
diameters of his disc, and the direction of each satellite's

motion noted, to avoid confusing the near and remote halves
of its orbit. The approximate periods here given will be of
use: I—1^d 18^h 28^m. II—3^d 13^h 15^m. III—7^d 3^h 43^m.
IV—16^d 16^h 32^m. Spots, as we shall presently see, may
cause this variable light: but a stranger source of anomaly has
been perceived,—the discs themselves do not always appear
of the same size, or form. Maraldi noticed the former fact
in 1707; Herschel I. 90 years afterwards, inferring also the
latter; and both have been since confirmed. Beer and Mädler,
Lassell, and Secchi have sometimes seen the disc of II larger
than I, and Lassell, and Secchi and his assistant, have distinctly seen that of III irregular and elliptical; and according
to the Roman observers, the ellipse does not always lie the
same way. Phænomena so minute hardly find a suitable
place in these pages, but they seem too singular to be omitted;
and in some cases possibly small instruments may just indicate
them; at least, with an inferior fluid achromatic reduced to
3 inches aperture I have sometimes noticed differences in the
size of the discs which I thought were not imaginary.

3. Their *Colours.* Different eyes and instruments have
here given different results. Herschel I. makes I and III
white, II bluish or ash-coloured, IV dusky and ruddy.
Beer and Mädler call I rather bluish, II and III yellowish,
IV always bluish. Secchi finds III sometimes whitish,
generally red. I have often thought IV ruddy. One unfavourable night of very thin white haze (1832, Oct. 8.)
I found the colours of II III and IV unusually contrasted,
though the discs were not well seen. Many times, both with
the fluid and the $5\frac{1}{2}$ feet achromatic, I have fancied IV, when
distant from Jupiter, and especially when viewed ob-
liquely, encompassed with a little scattered nebulous light.

It certainly is very unlike III when they are near together; but the difference perhaps arises only from colour.

4. Their *Eclipses.* The fading away or breaking forth of these little attendants, as they pass into or out of the great cone of shade which their monarch casts behind him for half a hundred millions of miles, is always interesting, and, when not too near the primary, within the reach of moderate instruments. The time must be taken from the almanac, long enough beforehand to be quietly prepared with the most suitable eye-piece, and a sharp look-out must be kept, as the diminution or increase of light, though not instantaneous, is speedy, especially with I and II. The planet had better be concealed behind a bar. Both Immersion and Emersion may be seen of III and IV, if not too near opposition; this can seldom be done with II, never with I, as the disc of the primary interferes.

5. Their *Occultations* by the globe of Jupiter: frequent, but not easily made out, nor interesting, as they shew no effect of the planet's atmosphere. Schumacher once saw a satellite hang on the limb, and seem to recede again, or make an indentation in it; but the eye is sometimes false from weary gazing in these cases, and such a solitary instance is worth recording and no more.

6. Their *Transits.* The most beautiful phænomena of this beautiful system; often recurring, and not too difficult for a moderate telescope: the first known instance having been seen as early as 1658, with one of the old unwieldy refractors, by its skilful maker Campani, and my fluid achromatic, 3 inches aperture, having often given me a pleasing view of the scene. When a satellite is seen rapidly approaching Jupiter, on the *f* side (that is, the side *following* as the object passes through

the field), a transit is inevitable : the satellite will glide on
to the disc like a brilliant bead, and remain visible from its
greater brightness for some distance, — according to South,
$\frac{1}{6}$ or $\frac{1}{8}$ of Jupiter's diameter, till it is lost in the luminous
background, to re-appear after a time, and pass off in the same
manner. But this is not all. An astronomer on Venus might
witness a similar transit of our Moon across the Earth at the
time of one of our solar eclipses, but he could scarcely if at
all perceive the black dot of shade which our attendant casts
upon us, as, from our comparative nearness to the Sun and the
breadth of his disc, the cone of lunar shadow tapers so rapidly that its end falls short of the Earth in annular eclipses,
and in total ones covers so small a spot on the surface that it
would be invisible at any considerable distance.* But
Jupiter is so much further from the Sun that the shadows of
his satellites form much longer cones, and falling but little
diminished upon his disc, traverse it as singular and conspicuous objects, perfectly round and as black as ink, at varying
distances p or f, (that is, preceding or following through the

---

* The greater axis of the elliptical dark spot which traversed England
(very nearly in the course of the annular eclipse of 1858) during the
great total eclipse, 1715, Apr. 22, is given at 150 miles, from Dartford to
Oswestry, in a curious " Description" or map of its path, by Halley, now
in my possession. It is accompanied by the following characteristic
notice:—" The like Eclipse having not for many Ages been seen in the
Southern Parts of Great Britain, I thought it not improper to give the
Publick an Account thereof, that the suddain darkness, wherin the
Starrs will be visible about the Sun, may give no surprize to the People,
who would, if unadvertized, be apt to look upon it as Ominous, and to
Interpret it as portending evill to our Sovereign Lord King George and
his Government, which God preserve. Hereby they will see that there
is nothing in it more than Natural, and no more than the necessary
result of the Motions of the Sun and Moon; And how well those are
understood will appear by this Eclipse."

field,) according to the relative positions of the Sun, Earth, and Jupiter. When not near opposition, the shadow, especially of III or IV, may be far within the disc, while the satellite shines out in the deep blue sky. Occasionally two of these total solar eclipses may be seen at once on Jupiter, and it will be interesting to mark their unequal velocities and distances from the satellites to which they belong. — Such is the regular mode of transit. But remarkable exceptions are not uncommon, owing to the variable brightness of the satellites, which sometimes cross the disc as dusky or even black specks; when deepest, almost like their shadows. This seems to have been first noticed by Cassini in 1666, but was more fully described by Maraldi in 1707, and referred to rotating or variable spots. After a long interval Schröter and Harding reobserved the phænomenon in 1796,* and perceived that the spots were sometimes only partial. More recently the great telescopes of Lassell, Bond, and Secchi have shewn that these little moons are liable to the formation of spots, just dark enough to be imperceptible in front of the fainter limb of Jupiter, but to start out rapidly in advancing upon the greater brightness of his centre, where they are generally dusky, irregular, and smaller than the shadow, proving their partial extent. I have sometimes seen them readily with my 5½ feet telescope. South, who once found with his large achromatic, two satellites of a light chocolate colour and their shadows on the disc at once, (which must have been, what he calls it, " a glorious view") says he never

* In 1785 and 1786 Schröter repeatedly saw round black spots traversing Jupiter rapidly, like shadows of satellites. It seems very unlikely that he should have been misled by these *dark transits*, yet that suspicion will arise. It is quite unaccountable that Herschel I. and Beer and Mädler should have passed by such obvious phænomena in silence.

saw one black : but at Cambridge U.S. 1848, Jan. 28, when the shadows of I and III were both in transit, III itself "was seen with the great refractor" * (Bond's) "under very beautiful definition, as a black spot between the two shadows, and not to be distinguished from them except by the place it occupied. It was smaller than its shadow in the proportion of 3 to 5, not duskish simply, but quite black like the shadows." Hartnup at Liverpool has also seen IV "only a few shades lighter than the shadow." Sometimes these spots have been so marked on III as actually to be visible to Lassell and Secchi, when it was shining freely in blue sky; the latter, who sees it reddish, and like the spotted face of Mars in a small glass, has even given drawings shewing them and their rapid motions in 1855. These are so curious and so little known in England that I have been induced to copy some of them, though most inaccessible to "common telescopes." Secchi would hence

Aug. 26. 19^h 20^m.   Aug. 26. 21^h 10^m.   Aug. 27.   Sept. 9.

infer a swift axial rotation : if so, these satellites must differ essentially from our Moon : that they do so to some extent is clear from the ephemeral nature of their obscurations, the same satellite often passing across Jupiter bright and dark at the interval of a single revolution in its orbit. An extract from my own observations, of a date long anterior to those of Lassell and Secchi, may be permitted as an encouragement to the possessors of ordinary instruments. "1835. Jan 26.

* 22$\frac{2}{3}$ feet focus, 15 inches aperture, by Merz, Frauenhofer's successor.

10^h 30^m. 250 [power of $5\frac{1}{2}$ feet achromatic]. The 4th [a little past inferior conjunction] was very pale, and rather ruddy; its disc not only smaller than that of the 1st or 2nd, but apparently imperfect as if spotted. Had the night been more favourable, this might have been a very interesting observation." Gruithuisen says that on two occasions he has seen the light of III reduced to a mere ring.

From this change in the aspect of a satellite in dark transit, Lassell justly infers a far greater difference of brightness between the centre and circumference of Jupiter than could have been suspected; a white disc would not be converted into a black one from mere contrast, unless the light of the background varied in a corresponding manner: and it is very strange that this variation is so little otherwise apparent to the eye. On optical grounds we should hence deduce a very smooth surface for Jupiter, and one not covered with vapour, since this, as we see in the dense white clouds of our own sky, would not grow less bright towards the edge. But there are other mysteries about this system. When Cassini once could not perceive the shadow of I, though it should have been on the planet, we may charge it upon the imperfection of the old refractors; yet they were of excellent workmanship, and Cassini was seldom wrong. But Lassell, a witness above all suspicion, has seen the shadow of IV very much larger than the satellite, apparently "twice as great in diameter," even *before* the latter had entered upon Jupiter, when it should have been enlarged by irradiation; and has found that the same shadow actually surpasses that of III, though the direct measures of their discs give an opposite result.* South

* Gruithuisen had noticed that the shadows were too large. Secchi saw this with II. An analogy—not an explanation—may be found in

many years ago published in one of the public journals a most interesting observation, which I greatly regret that I cannot recover, but I am confident as to its tenour, which was, that in his great achromatic he perceived each of two shadows of satellites on Jupiter to be attended by a faint duplicate by its side; traces of which could be just detected with a smaller telescope of (I believe) 5 feet. — But the most surprising is a phænomenon which requires, and possesses, the highest attestation. 1828, June 26, II, having fairly entered on Jupiter, was found 12 or 13^m afterwards *outside the limb*, where it remained visible for at least 4^m, and then suddenly vanished. The authority of such an observer as Smyth would alone have established this wonderful fact; but it was recorded by two other very competent witnesses, and (what is especially remarkable) at considerable distances, Maclear at 12 miles and Pearson at 35 miles from Smyth at Bedford. Explanation is here set at defiance; demonstrably neither in the atmosphere of the Earth nor Jupiter, where, and what could have been the cause? At present we can get no answer.*

———◆———

SATURN.

FORTUNATELY for the student, a common telescope will exhibit some part of the wonders of this superb planet, unparalleled in our own system, invisible elsewhere; and they whose ex-

the duration of lunar eclipses, which proves the shadow of the Earth to be larger than it should be from theory.

* Gambart says that at the immersion of I, 1823, Oct. 19, " le satellite a disparu et reparu plusieurs fois:" but in this case qu. as to the state of the air, or the eye?

pectations have not been unduly raised by designs from the
best telescopes will be delighted with the scene in their own.
The minuter details require in general great optical perfection;
but some of them may now and then be reached, though with
a feeble grasp, by ordinary instruments, and our readers may
have occasional access to more powerful means; it seems best
therefore to be circumstantial. We shall describe the Globe,
the Rings, and the Satellites, in succession.

The *Globe.* Though about 77,000 miles* in equatorial
diameter, second only to Jupiter, and nearly 10 times larger
than the Earth, this noble ball has so little density that it
would float like cork on water, which is about $\frac{1}{4}$ heavier, and
therefore, if any were found there, would sink to its centre;
while on the probable supposition that the density of the
globe decreases outwards, its surface-material must be lighter
still, and the conditions of existence there wholly unlike our
own. Day and night, clouds and winds, summer and winter,
the only steps in the analogy, leave an immense distance to
be overpassed; and the tenfold nearer approach of modern
astronomers has not bridged over the chasm which foiled the
enquiries of Huygens, and kept him still in ignorance of the
wonders beyond it.†

    1. Its *Ellipticity.* There is no doubt of this compression,
but much uncertainty as to its amount; the inclination of the
axis, corresponding with that of the ring, causes some diffi-

---

* Measurements differ a little; Hind's results have been adopted
throughout.

    † " Ad longinqua Saturni regna propius nunc quam antehac quis-
quam adivi, et usque eo progressus sum, ut vasti adeo itineris pars una
centesima tantummodo reliqua fuerit: quam si quo pacto superare
potuissem, quot, qualiaque . . narranda haberem!"

culty; the intersecting outlines of the ring occasion more, deceiving the eye so that at one time Herschel I. and others imagined that the ball was doubly flattened, at the equator as well as the poles. The best values of the compression vary between a little less than $\frac{1}{9}$ and a little less than $\frac{1}{12}$: at any rate it is very considerable. It is of course best seen during the disappearance of the ring.

2. Its *Excentricity*. Schwabe, though perhaps anticipated in perceiving that the globe is not accurately in the centre of the ring, first drew general attention to the fact, observed by him in 1827 with a $3\frac{1}{2}$ feet achromatic. It has been doubted, but seems established by measures which give the dark opening a very little smaller on W. than E. side : * the Roman observers considered it very evident, but rapidly variable, in 1842 and 1843 : I have sometimes fancied it visible with the $5\frac{1}{2}$ feet achromatic. It has been thought theoretically essential to the stability of the ring.

3. Its *Belts* and *Spots*. Darker and lighter bands diversify the globe, drawn lengthways by a swift rotation, changeable, and more varied in colour than those of Jupiter. Herschel I.'s "quintuple belt" of alternate light and shade is now seldom to be seen : usually a strong whitish band, the brightest part of the ball, surrounds the equator; on each side is a broad brown or somewhat ruddy zone, with many streaks and markings, generally parallel ; then a similar greenish or bluish grey region extending to the poles, which are capped by a darker patch of the same tint, with sometimes a pale central space. Strange to say, Beer and Mädler missed these belts, which Huygens discovered with his old refractor, and I have often seen.

* Gallet, the supposed discoverer, reverses this position.

—Spots are uncommon, and the instances easily enumerated : Herschel I., 1780 * — Schröter and Harding, 1796, 1797 — Schwabe, 1847 — Busch, a bright spot, 1848. Bond in 1854, and De La Rue in 1856, figure some darker patches in a belt. — From the unequal strength of parts of his quintuple belt Herschel I. found the rotation $10^h 29^m 17^s$. Schröter, from various spots, $11^h 51^m$, $11^h 40^m 30^s$, and rather more than 12^h. The precise period is probably not known. Another fact indicated an atmosphere to Herschel I., — the hanging of the satellites on the limb previous to passing behind it, the reverse of Jupiter. This appearance, unrecorded by others, should be carefully looked for towards the disappearance of the ring.

The *System of Rings*, so far as has been ascertained, is unique ; and the only level surface, or rather collection of level surfaces, that we know. From the combination of its inclined position, its parallelism with itself, and Saturn's revolution round the Sun, it will alternately shew each side to a distant spectator, and in two opposite points of the orbit come into an edgeways position and disappear. The whole revolution occupying nearly $29\frac{1}{2}$ years, all the phænomena of increasing and decreasing breadth will be gone through in something less than 15 years, for each side of the ring : and the edge-view having occurred in 1848, and the widest opening in 1856, the next disappearance will be in 1862, and the next full view (if the term may be applied when the breadth almost exactly equals half the length,) in 1870. To a fixed spectator in the Sun, the ring would vanish but once as it changed sides to him, from thinness alike and want of light;

* The object figured by Herschel I. in Phil. Trans. 1790, is called " a strong, dark spot near the margin of the disc," where even Bond's great achromatic fails to trace the belts, and where, on Jupiter, the darkest spots would be imperceptible through his atmosphere.

on the Earth, being neither in the centre of Saturn's orbit, nor
at rest, we may find it disappear in four positions — either
edgeways to Sun and Earth, — edgeways to Sun but not to
Earth, when it will be too faint, each side having only a very
little horizontal light, — edgeways to Earth but not to Sun,
when, though enlightened, it will be too narrow to be visible,—
or edgeways to neither, if its dark side is towards Earth: the motion of Saturn however carries it before very long through all
these points, and the previously hidden side comes effectually
into view.   We must now consider it as it was in 1856, when
the ring at its greatest width projected beyond each pole of the
globe: in this position it is brought out by a very small power,
the blue sky being visible on each side through the "ansæ"
(or *handles*), as the semi-ovals of light are termed into which
the ring is projected.   Galileo's telescope, though it magnified
upwards of 30 times, could not define it thus, and he imagined
that he saw a large globe between two smaller ones: then
followed a number of queer-shaped misrepresentations by
various hands, till in 1659 Christian Huygens made for himself a refractor of 23 feet focus and $2\frac{1}{3}$ inches aperture, which,
bearing a power of 100, solved the mystery.   His exulting
father celebrated the discovery by a copy of verses, closed by
a passage worthy of preservation, in which he thus anticipates
for his son a fame as lasting as the firmament;

> Gloria sideribus quam convenit esse coævam,
> Et tantum cœlo commoriente mori.

The telescope now improved rapidly, and the cumbrous and
troublesome refractor of those days was often exceedingly good
for the kind of thing.   1665, Oct. 13, two brothers at Mine-
head with a 38 feet telescope discovered the principal division

K

in the ring, * which ought to be called Ball's, after their name : it was rediscovered by Cassini in 1675, one of whose refractors, 20 feet long, and not quite $2\frac{1}{2}$ inches aperture, bore a power of 90. A much lower power would not shew it all round, as Cassini seems to have seen it, even in a modern achromatic. But we must proceed to details, adopting Otto Struve's designation of the three principal rings, by the letters **A**, **B**, **C**.

1. The *Outermost Ring* (**A**), about 172,000 miles outside measure, was perceived by Cassini, as early as 1675, to be considerably fainter than the outer part of its neighbour ; a small telescope will readily shew the difference. There seems reason to believe that its structure is either always, or temporarily, multiple. Short in the last, Kater in this century saw, on rare occasions, several concentric fine lines upon its surface. The strongest of these, named after Encke, who first drew attention to it in 1837, is perceived from time to time rather outside the middle of the ring. Jacob sees it constantly at Madras, and P. Smyth found it very distinct in the pure sky of the Peak of Teneriffe ; Lassell and Secchi find only a dusky mark like a pencil line, Dawes and De La Rue consider it a black division.† Immediately within it, the latter observer represents a brighter streak in the exquisite drawings taken in 1852 and 1856, with his superb reflector of 10 feet focus and 13 inches aperture. Secchi finds another dark marking interior to the first. Dawes, and Coolidge, Bond's assistant, have seen a

* Humboldt was not aware of this fact.

† In 1851 and 1852 Dawes saw this line when Bond, Struve, and Lassell, with three of the finest telescopes in existence, missed it. It was subsequently seen dusky in Bond's. As recently as 1858, Apr. 17, during an unusual display of Saturn, Lassell could only detect a very slight shade.

greater brightness at the inner edge, Lassell (at Malta) at both
edges of this ring.   Common telescopes of course break down
here, yet I have once or twice suspected a streaky aspect with
the $5\frac{1}{2}$ feet achromatic : but it is very strange that these markings have escaped many of the best telescopes, especially
directed to Saturn, — those of the two Herschels, * Schröter,
and Struve at Dorpat.   Possibly they may be subject to
obscuration or change.

2. The *Ring* **b** has a luminous region at its outer border,
sometimes the brightest part of the whole system, though according to Schwabe variable, a circumstance which I have also
noticed, finding it more conspicuous in 1853 and 1855 than in
1856 and 1857 : the inner edge is very obscure ; and from one
rim to the other there is an increasing shade, formerly considered regular, now found by Lassell, Dawes, and Secchi, to
consist of 4 or 5 concentric and deepening bands, compared
by Lassell to the steps of an amphitheatre ; each may be a
separate ring, since fine black lines were traced here by Encke
and De-Vico,† and Bond's telescope shewed markings like
narrow waves of light and darkness.   De La Rue, and Dawes
in another year, have seen a lighter central band.   The innermost edge is sometimes brightened up — in other seasons it
fades.   I saw it in 1853 without any previous knowledge, on
one ansa, with only $3\frac{7}{10}$ inches aperture : in 1856 it was less
visible, and went beyond my range in 1857.

3. The *Ring* **c**, the crape or gauze veil, is one of the

* In an admirably perfect view of Saturn by Herschel II., 1830, Apr.
4, with his 20 feet reflector, $18\frac{3}{4}$ inches aperture, no subdivision could be
traced " with all possible attention and with all powers and apertures."

† Herschel I. saw such a line in June, 1780, for a few days on one
ansa only.   De-Vico's observations are not always satisfactory.

greatest marvels of our day. How it could have escaped so long, while far minuter details were commonly seen, is a mystery indeed. Schröter in 1796, with his great reflector, 26 English feet focus, 19 inches aperture, particularly examined the space on each side of the ball, and found it uniformly dark; if anything, darker than the sky.* Our great observers the two Herschels never perceived it. Struve measured Saturn repeatedly in 1826 with the superb Dorpat achromatic, 14 feet focus, 9·43 inches aperture, but missed it, though he saw the inner edge of **B** very feebly defined. In 1828 a fine Cauchoix achromatic of $6\frac{1}{3}$ inches aperture having been placed in the Roman observatory, it was seen, as an old assistant informed Secchi, both in the ansæ and across the ball, yet, strange to say, and little to the credit of Roman science at that time, *no notice whatever was taken of it.* Ten years later, in 1838, the matter fell into worthier hands; — Galle of Berlin, the optical discoverer of the planet Neptune, with a glass the counterpart of that at Dorpat, detected and measured it very accurately, and published his observations; yet they somehow went on the shelf till in Nov., 1850, Bond in America, and our own Dawes, had each the honour of an independent and original discovery; and now everybody may find it with a sufficient aperture: it has been well seen with a $3\frac{3}{8}$ inch achromatic by Ross; I saw faint traces of it with my $3\frac{7}{10}$ inches; anything larger, if good, is sure to bring it out.— Where was it, for so many years? Probably some change may have rendered the part projected on the sky more luminous; but it is not a new formation, for where it crosses the ball it has been seen ever since Campani's time, 1664: and it was even re-

---

* Bond's telescope shews the same appearance between **C** and the ball.

peatedly noticed that its outline did not suit the perspective of
a belt on Saturn, as it was supposed to be, or that of a shadow
thrown by the ring: on this bright background it is so easy
that I seen it with $2\frac{1}{4}$ inches aperture. It reaches rather
more than half way from the edge of **B** to Saturn: at its
first discovery Dawes, and Otto Struve at Poulkova (the Czar's
observatory) with the superlative Frauenhofer achromatic,
22 feet 4 inches focus, $14\frac{9}{10}$ inches aperture, considered it to
be divided in two by a dark line; but this has not been seen
since; nor is it certain whether there is a division between it
and **B**, or whether it is fainter towards its inner edge. Its generally slaty hue has been at times found different on the two sides
of the planet, reddish and bluish; and a repeated interchange
of these tints, noticed by Lassell, may indicate rotation; its
width in the two ansæ has been once or twice seen different;
and its narrowness in front of the ball has induced an idea that
it is not in the same plane with the rings **A** and **B**. Lassell and
Jacob have discovered the very curious fact of its partial
transparency, permitting the limbs of Saturn to be traced
through it. Possibly a similar material may fill Ball's division,
as this has been sometimes seen not quite black, and Jacob
has even followed the shadow of Saturn across it. Many other
slight variations lead to the enquiry whether this grand system
is really permanent in its detail. Secchi thinks the bright
rings may be clouds in an imperfectly transparent atmosphere,[*]
which is visible in the ring **C** and Ball's division; and the
American mathematician Pierce considers it demonstrable that
the ring is not solid, but may be a stream or streams of a fluid

---

[*] The ring was suspected to have an atmosphere by Herschel I. from
the apparent threading of the minutest satellites upon it when reduced to
extreme thinness.

rather denser than water, adding in the spirit of a true philoso-
pher, " Man's speculations should be subdued from all rashness
and extravagance in the immediate presence of the Creator."
The transit of the rings over a considerable star would perhaps give us some further information; but such events are very
infrequent. Whiston states that Dr. Samuel Clarke's father
once saw a star in the sky between the ring and Saturn, but nothing of the kind has been since recorded. Dawes alone has
seen, but under unfavourable circumstances, a star between 8
and 9 mag. pass behind the outer edge of **A.** In such a case,
all attention should be given to the visibility of the star through
at least Ball's division, and the dark ring. The shadow of the
planet upon the rings is readily seen, except near the opposition, cutting off one of the four arms of the ansæ from apparent
contact with the ball: I have perceived it with $2\frac{1}{4}$ inches
aperture. The outline of this shadow has often been found
curved the wrong way for its perspective, or deviating on ring
A from its line on **B** ;* something of this kind has been noticed
ever since Cassini's time, and I have made it out with the
$5\frac{1}{2}$ feet achromatic : the shadow of the ring on the ball has also
been observed notched by Schröter and Lassell. Hence an
uneven surface would be inferred, and Secchi finds that a reversed curvature would result from a slight convexity in the
ring ; but in each case we are at once confronted by its total
or nearly total disappearance when viewed edgeways, seeming
to demonstrate absolute flatness : it cannot however be quite
symmetrical, for not only do the opened ansæ sometimes shew
an unequal breadth of markings E. and W., but about the
time of their vanishing or reappearance they have been re-

* Coolidge with Bond's telescope in 1856 saw it strangely flattened,
with a deep notch, and it was unequal in depth of shade.

peatedly seen, ever since 1714, variable and unlike in length and thickness. When the light of the ring has been reduced to a thread, knots of greater brightness have often been noticed upon it. From the motion of some of these in 1789, Herschel I. deduced a rotation in 10^h 32^m 15^s, but this is unconfirmed. Schröter and Harding in 1802 and 1803, Schwabe in 1833 and 1848, and Bond in 1848 found them all immoveable.* Schröter supposed that they were mountains; Olbers, more probably, that they were the perspective aspect of such parts of the ring as shew the greatest breadth of light. Bond finds this insufficient, as they are not lost when the dark side is turned to us, and he thinks that they may be explained by the reflection of light from the interior edges of the rings. Yet many observations indicate too much irregularity to be so readily accounted for.—When it stands edgeways, or turns its dark side to us slightly, the whole system totally vanishes in ordinary instruments. Herschel I. seems to have kept hold of it, and Bond did not quite lose it in 1848, though it was interrupted in places. The thinness of this enormous plane is literally matter of astonishment; Herschel II. estimated it at 250 miles. Bond, who allowed it but 0·01″, at 40 miles, that is, $\frac{1}{4800}$ of its diameter! In 1848 Dawes saw traces of the dark side when turned to us, of a deep coppery tinge. The disappearance in 1862 will be looked for with especial interest; and the possessors of ordinary instruments need not despair; the slender line of light is a beautiful spectacle; and my fluid achromatic of 4 inches aperture, but of which much of the light went the wrong way, lost it in 1833 only one day sooner

* De-Vico saw on several successive nights in 1840 a fixed bright point attached to the open ansa. This, with the instrument and eye of his successor Secchi, would have negatived a rotation at once.

K 4

than Smyth's great achromatic of nearly 6 inches, and but four days sooner than Herschel II.'s 18 inch reflector.* Saturn at such times shews usually a dark and very visible streak across his face, which Dawes has divided lengthways, — the unenlightened side of the ring, and its shadow. Schwabe and De-Vico have then thought the belts not exactly parallel to the ring.

The *Satellites.* The very confused way in which the members of this numerous retinue were formerly designated, by numbers counted in two different directions, has been superseded by Herschel II.'s ingenious nomenclature, in which the names reckoned towards Saturn are easily remembered in a Latin pentameter and a half —

> Iapetus, Titan, Rhea, Dione, Tethys,
> Enceladus, Mimas.

This arrangement has been since disturbed by the simultaneous discovery in 1848 by Lassell and Bond of an 8th, Hyperion, extremely faint, between Iapetus and Titan. — The lesser of them belong only to the highest instruments : the innermost has never been seen certainly with anything smaller than a 6½ inch object-glass, and then only by the perfect vision of Dawes, or in the purer skies of Paris, Rome, and Madras : the next is scarcely easier. But Titan, the leader, is very conspicuous, being equal to an 8 mag. star, so that it was discovered by Huygens in 1655, in which year it was also seen, though not suspected to be a satellite, in England by Sir Paul Neil and Sir Christopher Wren. I have seen it very plainly with a 2¼ inches object-glass when long past opposition, and I believe I have perceived it with a much smaller aperture. In fact

---

* Lardner says, Herschel II. did *not* lose it : not a solitary instance of inaccuracy.

this little point of light has in large telescopes a disc of about
0·75″, and probably ranks in size between Mercury and Mars.
Its shadow has been watched by Herschel I. in transit across the
disc of Saturn; Gruithuisen alone, if we may credit him, has
been equally fortunate.* — Iapetus is not small, though probably from a spotted surface and rotation like our Moon, not
always equally visible, being much brightest at his W. elongation.†—The next three are more minute.‡ Kitchiner states,
from a friend, that by hiding the planet, they have been seen
with $2\frac{7}{10}$ inches aperture; it is more certain that $3\frac{3}{4}$ inches
have shewn them. I believe I once saw them all with 4 in-
ches (not good). They must not however be near the
powerful light of the planet. Schröter also found them
variable like Iapetus, but brightest towards E. instead of W.;
and the inclination of their orbits, nearly similar to that of
the ring, by spreading them over the sky, increases the difficulty of distinguishing them from small stars in one night's
observation. To a spectator placed on Mimas, revolving in
less than 23^h at a distance of only 32,000 miles from the edge
of **A**, the whole system of rings and the included globe would
float before the eye in such a spectacle of grandeur and beauty
as the imagination is wholly unequal to conceive.

* He states that at the same time he saw Enceladus and Mimas,
though with only $4\frac{1}{4}$ inches aperture.

† A confusion exists in some astronomical works with respect to the
relative size and variable light of this satellite and Titan.

‡ That Cassini with his inferior means should have not only discovered
them, but traced their periods, reflects great credit upon his ability; it
is gratifying to be able to subjoin the striking encomium of his Eloge:
"Les cieux qui racontent la gloire de leur Créateur, n'en avoient jamais
plus parlé à personne qu'à lui, et n'avoient jamais mieux persuadé."

URANUS AND NEPTUNE.

THESE planets may be reached, but to no great purpose, with ordinary means. URANUS, being visible in clear weather to the naked eye, will be easily caught up in the finder by the help of the Almanacs before mentioned, and will be large and planetary-looking in the telescope; its disc indeed subtends 4″, but I never found the light of $3\frac{7}{10}$ inches sufficient to define it perfectly : with Lawson's 7 inch object-glass, bequeathed to the Greenwich Naval School, I have seen it beautifully, as a little Moon : no one has made out much more : Mädler has seen it flattened : Lassell with his 2 feet mirror, and in the sky of Malta, once thought it possible there might be a spot : the two rings suspected by Herschel I. have long been abandoned, and Lassell has reduced the satellites to 4, which even the 6 inch object-glass of Smyth was incapable of reaching.

NEPTUNE may be found by the Nautical Almanac; but will hardly repay the search ; I have several times seen him, dull and ill-defined, with my $5\frac{1}{2}$ feet achromatic ; his satellite of course no common telescope will touch. But who can say how grand a spectacle he might present on a nearer approach? or that he is the most distant planet that obeys our central Sun? or what unexplored wonders may lie in still remoter space? " Quis unquam exhaustas dixerit cœli copias? " * Another generation will probably open up fresh marvels, or prove to us that, as far as the dominion of our own Sun is concerned, we have reached the boundary of our knowledge.

* Bianchini.

COMETS.

Thou wondrous orb, that o'er the northern sky
Hold'st thy unwonted course with awful blaze !
Unlike those heavenly lamps, whose steady light
Has cheer'd the sons of earth from age to age,
Thou, stranger, bursting from the realms of space
In radiant glory, through the silent night,
Thy tresses streaming like the golden hair
Of Atalanta, or that beauteous maid
Pursued by Phœbus, upward shalt invite
Many a dull brow unus'd on heav'n to turn,
And many a bosom rend with deep alarm.

Where is thy track throughout the vast expanse ?
Still onward hast thou urged thy bold career,
From that first hour when the Creator's hand
Impell'd thy fire along the fields of light,
Nor ever yet arriv'd within the verge
Of mortal ken, nor drank the distant beams
Of our inferior sun, unask'd by thee
To guide thee harmless on thy rapid way ?

REV. JOHN WEBB (1811).

WHEN Kepler stated his belief, not merely that comets inhabited the æther as fishes the ocean, but that the ocean was not fuller of fishes than the æther of comets,* his contemporaries probably amused themselves with his luxuriant fancy. Yet Kepler was not far wrong as regards number; for it is now believed that upwards of 4,000 have approached the Sun within the orbit of Mars during the Christian æra. A twelve-month never passes by, in these days of telescopic activity, without the announcement of several; 1858 produced 8, and sometimes 3 and even 4 (as in Feb. 1845) have been in sight at once. The generality are faint, and so much alike,

* Nec minus ætherem cometis refertum esse puto, quam oceanum piscibus. Quod autem rari apparent nobis, ingens ætheris vastitas in causâ est.

at least in ordinary telescopes, that they offer little attraction beyond the curiosity of watching their progress: this, where there are stars for sky-marks, may be sometimes traced from hour to hour, and, in connection with the enormous distances traversed, gives a grand idea of the majestic movements of the universe. But larger comets are often equally imposing to the naked eye and marvellous in the telescope; of which we have recently had a splendid instance in " the Donati," or in astronomical language, Comet V, 1858, that is, the 5th in order of that year. It would lead us into too wide a field, as well as one already cultivated by abler hands,* were we to describe separately the more remarkable of these bodies; it will be better therefore to sketch a general outline, which may serve to direct our expectations in future. First, however, we will give directions for finding them, when invisible to the naked eye.

If a comet cannot be readily perceived by moving about, over the proper quarter of the sky, a hand-telescope with a large field (an opera-glass may answer well), it is too faint, and we must take our larger instrument and " sweep " for it, thus: ascertain its probable position — not, of course, that of the date of its discovery, but that of the present time, if the announcement gives a clue to it: put in the lowest eye-piece, point the telescope some way below, and on one side of, the supposed place of the comet, and gently move it horizontally right or left, as the case may be, till it points about as much on the other side. When your first " sweep " has been thus

* See especially Hind's excellent treatise " The Comets " (1852) and Bond's admirable " Account of Donati's Comet," Cambridge, U. S., 1858. Pingré consulted 616 authors for his Cométographie, and there were, in 1841, 382 works on the subject in the Czar's observatory at Poulkova.

completed, raise the telescope vertically half the breadth of the field, or more (as a faint object too near the top or bottom might be missed), and sweep back again the same distance the reverse way, ending pretty nearly where you began, but of course half a field higher; raise again another half field, and sweep on, till either the comet sails across the field, or you are evidently pointing much too high for it: in which case you may conclude that it is too faint for you, or has gone away so far as to make further search useless. The movement may of course as well be begun above as below the supposed place of the comet. This process of sweeping is equally available for finding Mercury or Venus in the day-time, or minute stars or nebulæ at night; the chief difficulty lies in so proportioning the vertical movement that the object shall not escape between the sweeps; this is easy where there is rackwork, as any terrestrial object will serve to ascertain the breadth of the field, and the suitable number of turns of the handle; with a plain stand there is little trouble in it where stars occur frequently to guide the displacement of the field; but by day, or in twilight, or in barren tracts of sky, sweeping without rackwork is less certain, and may require to be repeated.

The light of most comets is too faint for high powers, and requires a contrast with the dark sky which their small fields do not admit: to get the whole extent of head or tail the lowest power should be chosen, and for the tail the naked eye will be more effective than any glass, except the appropriate instrument called a Comet-finder: high powers may be used for the details of the nucleus, but the contraction of the field and the want of contrast must always be allowed for. Small comets are frequently nothing but luminous mists without trains, sometimes without central condensation; — as we ascend in the scale, nuclei, trains, envelopes, and various anomalous appearances succeed, depending, it is probable, upon diversities in the materials of the comets themselves, as well as on their degree of approach to the Sun ;—even Venus was thought to have influenced the aspect of "the Donati," which passed very near her. In common language, a comet consists of *head* and *tail* ; but we have to make a telescopic analysis, and shall begin with

1. The *Nucleus :* the most luminous part, occupying in a general sense the centre of the head. It is sometimes absent in telescopic comets, which are mere fogs, permitting the minutest stars, even such as would be effaced by the slightest terrestrial mist, to be seen through the very centre ;—thus Herschel II. saw a group of 16 and 17 mag. stars through the heart of the Comet of Biela in 1832 : from this diffused state they present every stage of condensation, including sometimes a sparkling or granulated appearance, up to the aspect of a star, which has sometimes, as in 1744, equalled Venus in brightness, and been visible to the naked eye in broad noon-day, or as in 1843, blazed out yet more splendidly like a bright white cloud, close to the glowing meridian Sun.* High powers, however, usually dissolve any apparent solidity, and different instruments will give (as was the case with Donati's Comet) very different sizes to what looks like a planetary disc, throwing much doubt upon its reality : sometimes a very minute point (as Herschel I.

* Other comets visible by day (omitting some questionable ones):— A.C. 43, the star of "the mightiest Julius," which appeared during the games held soon after his assassination, and gave rise to the star on the forehead in his coins and statues;—1106, 1402, 1577 (Tycho's great comet), 1618. That of Feb. 1847 was seen by Hind, and that of Aug. 1853 by Hartnup and Schmidt very near the Sun ; but in each case with the telescope ; in which they shewed planetary discs. Donati's comet was not visible, except with very powerful instruments, by day.

found in 1807 and 1811) holds out against any magnifying; and this — like broader nuclei — is not always central in the head. In some comets (1618, 1652, 1661, 1707) the nucleus seems to have been composed of separate masses, — a marvellous structure;*—would that we could study it with modern advantages! Had "the Donati's" nucleus passed about 20′ further N., it would have gone right over Arcturus — and the curious question as to solidity might have received some kind of reply. But no such instance of transit over a brilliant star is on record, and some minor examples are not conclusive. Had a defined phasis ever shewn itself, solidity, or at least considerable density, might be inferred, but this has never been satisfactorily the case ; the retroverted form often assumed by luminous sectors throws a doubt upon the observations; and the shadow sometimes ascribed to the nucleus is equally ambiguous : the darkness, however, close behind it in 1858 seemed to prove that the nucleus was not permeable to the solar energy, in whatever way exerted, and that it had no rotation upon an axis.

2. The *Coma* is the sphere of mist around the nucleus, forming in popular language the *head;* this is sometimes of considerable extent—2° 40′, or 5 times the Moon's diameter, in 1770, when the nearest recorded approach to the Earth took place, of 1½ million of miles. The coma fades away into the surrounding sky : its denser part, if distinguishable by any set-off or outline, is called

3. The *Envelope.* This is an interior and brighter layer

---

* There is a strange story of visible internal movement among these subordinate nuclei in 1618 — as if of coals stirred in a fire. This seems like a bad telescope or an alarmed observer — yet where all is mystery, it is not wise to be incredulous.

of mist, suspended as an atmosphere around the nucleus, at least where exposed to the Sun : in Donati's comet some observers (amongst them, myself) carried it a good way round the back of the nucleus ; but usually it turns straight off on each side to form the commencement of the tail : within it is sometimes a darker narrow band, separating it from another interior and brighter envelope ; in 1858 no fewer than four such dark spaces at once were sometimes shewn by the great achromatic at Cambridge, U.S., indicating the consecutive rising of five shining waves, which had spread themselves in succession outwards from the nucleus ; the latter seeming to diminish in size and brightness after each of these emissions of luminosity. Mädler at Dorpat noticed an actual increase of one of these envelopes from $18''$ to $27''$ in 2^h. In 1811 the whole coma and envelope were raised in one parabolic-shaped mass from the nucleus, which was surrounded by clear dark sky on every side ; when the extreme diameter of the luminous cloud amounted, according to Schröter, to 947,000 miles, considerably exceeding the bulk of the Sun, and almost doubling the Moon's orbit. As this comet receded, the dark space became indistinct, and the envelope sank finally down upon the nucleus. Envelopes sometimes contain dark spots and streaks (1858), and more frequently brushes or fans of light, where the shining matter seems to stream out from circumscribed portions of the nucleus ; these are rapidly variable, and in Halley's comet at its return in 1835 shewed such a libration from side to side, traceable from hour to hour, that Bessel inferred some powerful polar force unconnected with gravity. Such a swinging was less distinctly recognised in 1858. The envelope and coma together form the origin of

4. The *Tail*. This, when not greatly foreshortened, appears

as a long cone (or including the head, paraboloid) widening usually as it advances, and shewing a hollow structure
by an interior darkness. I have ventured however to doubt
the adequateness of the explanation; since the perspective of

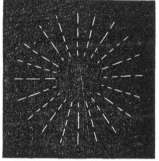

1858 required a very different
proportion either of the
breadth, or the *intensity* of the
dark space : a combination of
a hollow interior with a radi-
ated structure, something like
the diagram, would I believe
better agree with observation,
and such a structure would
exactly harmonize with the
most probable theory of comets—that of Olbers—which
ascribes to them a polar condition excited by the Sun, and
similar to electricity or magnetism. This interval in the
Donati was very dark near the nucleus; Schwabe found it
darker than the twilight sky, and Hartnup considered it *real
shadow* : it was however probably not so, or at least only in
part, as in 1811 such a vacancy surrounded the nucleus on
every side. The division into two streams effected by it
was seen with the naked eye in the great comet of 1577, which
astonished Tycho at his fish-pond before sunset, and it has been
frequently noticed since the discovery of the telescope; in
1843 it was absent; and a brighter ray filled the middle when
Halley's comet in 1835 was withdrawing from the Sun. The
preceding stream according to the comet's motion is usually
brighter and sharper defined, as well as curved backwards.
Kepler noticed this in 1618, and compared it in his own graphic
way to the appearance of a heap of corn swept over by the

wind : and no one will forget it in the shape of the Donati.* It
has been usually referred to motion through a resisting medium,
but Pape, from an elaborate discussion of our recent visitor,
finds that the curvature may be explained by the combination
of the comet's motion with a repellent power in the Sun.
The axis of the tail was calculated by him to differ 6° 18′
from exact opposition to the Sun, being left a little behind;
—a deviation of which there have been previous examples.
The length of the tail is sometimes enormous. According to
Boguslawski, that of the wonderful comet which in 1843
nearly grazed the surface of the Sun, and, as Secchi expresses
it, issued from it like a dart of light, was cast away to the
almost incredible extent of 581 millions of miles—more than
six times the distance of the Earth from the Sun, and crossing
the orbit of Jupiter : and this astounding stream, the longest
object, as far as our senses could reach, in all space, must have
been shot out with an equally inconceivable velocity after
passing the perihelion; for, but a short time before, the head
was on the other side of the Sun, half round which it had
been whirled in 1½ʰ or 2ʰ, and it could not possibly under
such circumstances have reversed the whole direction of the tail.
Here we seem to have before us one of the greatest marvels
in the universe.—Minor trains are occasionally seen; in some
cases (1769, 1811) the prolongations of an outer envelope;
in others apparently separate branches (1806, 1843); six of
them, if the observation is trustworthy, formed a glorious fan
in 1744. A straight, long, narrow, and very faint ray of this

* Towards the close of its appearance, the American observers
noticed a reversal of the brightness and definition of the sides of this
comet's tail. They also found, contrary to the common opinion, that
the plane of its curvature was not identical with that of its motion :
how this can be reconciled with Pape's theory does not appear.

kind was seen by several observers steadily preceding the great curved tail in 1858, and directly opposite to the Sun. From certain periodical returns of such streams Dunlop in 1825 inferred a rotation of the tail in $20\frac{1}{2}^h$, and similar appearances were noticed in 1769 and 1811, but the explanation seems a precarious one, from the immense velocity required, and the feeble mutual gravitation of the revolving particles. An additional stream has sometimes, though rarely, been directed *towards* the Sun; this extraordinary phænomenon, called an *anomalous tail*, was noticed in 1824, when it was longer and brighter, though narrower and more tapering, than the usual one; subsequently it was found that the grand comet of 1680—Newton's comet—had left a similar glowing wake upon the æther; and it has since been noticed in 1845, 1848 (Encke's), 1850, and 1851.* The principal tail in 1858 was seen by the Americans crossed obliquely by a number of brighter bands, like auroral streamers, diverging from a point between the nucleus and the Sun.

A few other details should be noticed. *Coruscations* or flashings have been often remarked in the tail: that of 1556 was said to waver like the flame of a torch in the wind; and numerous other instances might be given, before and since the use of telescopes. The accurate Hooke took many precautions before he satisfied himself of their reality in 1680 and 1682, and Schröter stoutly maintained their existence in 1807, referring them to electricity or some analogous cause : others negative them as too rapid for the progressive motion of the light by which we should see them—an objection, however, applying only to foreshortened tails,—and treat them

* One of the three short rays, besides the tail, seen in 1577, seems to have been of this character.

as illusions depending on our own atmosphere, or the uncertainty of weary sight. Were they established, they would go some way to decide the disputed question as to native or borrowed illumination, which the usual want of phases leaves open. Optical experiments, in shewing that a part of the light is reflected, do not disprove the co-existence of original light, which would rather follow from their transparency to minute stars, as a medium transmitting such delicate rays would scarcely detain and reflect those of the Sun. " Omnia incerta ratione, et in naturæ majestate abdita."*

The *colours* of comets differ ; a wide margin must be left for the superstition of the ancients, who dreaded the herald of disaster, " terris mutantem regna cometen," † and held it as a malignant genius that

> —— " from his horrid hair
> Shakes pestilence and war."

Those who noticed the fiery hue of the sabre-like comet that passed to the E. at the commencement of the Crimean war will understand with what feelings our fathers would have gazed upon it, like the astonished spectators of the comet of the Norman Conquest, represented in the Bayeux tapestry with the inscription " isti mirantur stellam :" nor will the exaggeration of those beautiful lines seem unnatural,

> ——" liquida si quando nocte cometæ
> Sanguinei lugubre rubent."‡

It is however certain, both from the Chinese and modern observations, that there is much general difference in respect to colour, and Herschel I. even found the envelope of the

* Seneca. † Lucan. ‡ Virgil.

same comet (1811) yellowish, the nucleus bluish-green; an indication perhaps of different materials, from which also branching and anomalous tails may possibly take their rise. Of the nature of these materials we are utterly ignorant, though it is evident that they are discontinuous, like clouds of mist or dust, since, were they gaseous masses, they would displace by refraction the stars over which they pass: that the Sun induces in them a polar force superior to that of gravitation — the hypothesis, in the main, of Olbers — is most probable from the form of the envelope and tail, as if repelled alike from the nucleus and Sun: and that a certain portion of the rarer particles is dissipated during the perihelion passage is evident from the very aspect of a tail such as that of 1858, whose restoration to the nucleus is more inconceivable than the return of the steam to a railway engine at its fullest speed. And in this, and in the destructive resistance which one comet (that of Encke) is known, and all may be presumed, to suffer, from the denser æther around the Sun, we find a striking indication that our system was not made to be eternal. The perfect balance of its construction might at first lead to another impression, and seem to countenance the old objection, that "all things continue as they were from the beginning of the creation." But here is evidence to the contrary — a slight but decisive symptom that — " they all shall wax old as doth a garment " — that " they shall be changed." Whether a similar cause may have been acting upon the comet of Biela, it is not easy to say; but its separation into two parts is one of the marvels of modern astronomy. Single in all previous observed returns, (those of 1772, 1805, 1826, 1832,) in 1846 it became elongated, and then threw off a portion which increased till it rivalled and even for a short time surpassed its parent, each having at one period a starry nucleus and short tail, while they were connected by faint streams of light: and thus they continued in sight for more than three months, keeping a distance of something more than 150,000 miles, the companion being the first to vanish. When next seen by Secchi in 1852, the distance was more than eight times greater, being 2^m in RA, and $30'$ in Declination : and now they are probably independent bodies: the return of this comet during the present season is therefore looked for with more than common interest.

PART III.

THE STARRY HEAVENS.

"Lift up your eyes on high, and behold Who hath created these things, that bringeth out their host by number: He calleth them all by names by the greatness of His might, for that He is strong in power; not one faileth."

ISAIAH, xl. 26.

IF the Solar System had comprised in itself the whole material creation, it would alone have abundantly sufficed to declare the glory of GOD, and in our brief review of its greatness and its wonder we have seen enough to awaken the most impressive thoughts of His power and wisdom. But that system is but as a single drop in the ocean. What boundary may be set to creation we know not, but we can trace it far enough to perceive that, as far as our senses are concerned, it cannot be distinguished from absolute infinity: and in leaving our Sun and his attendants in the background, we are only approaching more amazing regions, and fresh scenes will open upon us of inexpressible and awful grandeur. We are now to contemplate not one Sun, but thousands and myriads: — not a planetary system of subordinate globes, but aggregations of Suns; — pairs, groups, galaxies of Suns — "the host of heaven," — all independent in unborrowed splendour, yet many evidently, and all by clear implication, bound together by the same universal law which keeps the pebble in its place upon the surface of the earth, and guides the falling drop of the shower, or the mist of the cataract. Many of these Suns may probably be smaller or dimmer than our own, yet others unquestionably far surpass his splendour; while as to distance, their remoteness is so inconceivable that light itself, flying with a speed which would encircle the Earth thrice in one second, only shews them to us as they formerly were, — some, years, — others, centuries, — others perhaps whole ages back, even in the first dawn of creation. Here is indeed a field where enterprize cannot be thrown away, nor perseverance fail of its reward.

We must, however, remember that, though they are Suns which we are contemplating, and though the mere aspect of some of them in a large telescope well bears out the assertion,* yet a great proportion are diminished by distance to the minutest points of light, and can only be distinctly seen under favourable circumstances. We cannot therefore expect unvaried success; in fact the more delicate objects of stellar astronomy are not only among the severest tests of a telescope, but are peculiarly liable to be affected by atmospheric indistinctness, and require the most propitious skies. The cautions suggested in Part I. will be especially applicable here.

An original investigation of all the objects worthy of notice, even in a popular sense, in the starry heavens, would have been the attempt of a life-time, rather than of such occasional

---

* The approach of Sirius to the field of Herschel I.'s 40 feet reflector is said to have been ushered in by a dawning light, and its actual entrance to have been almost intolerable to the eye: yet the 4 feet speculum was by no means good. What must have been the blaze of this star in the Earl of Rosse's telescope, with a speculum of 6 feet, of admirable quality!

hours of leisure as I could command; an unverified selection, on paper, from a standard list formed with a very different design, would have been an easy, but not a satisfactory task : a middle course has therefore been preferred. All such of the 850 Double Stars and Nebulæ of Vice-Admiral Smyth's Bedford Catalogue as my 5½ feet achromatic could be expected to reach, were examined in succession, and those only retained for our purpose which seemed to possess sufficient general as well as scientific interest, and might serve as specimens of the surrounding profusion : and as in such a review a number of other objects, beautiful to the popular eye, though unimportant perhaps to the professed astronomer, presented themselves unsought, many of these have been added to the list, identified, where practicable, from Struve's "Mensuræ Micrometricæ," and distinguished by brackets : a few which I have not seen are included in parentheses. To these, the attentive student will be continually gathering fresh groups and combinations, especially from the crowded fields of the Galaxy; — a very little personal experience will convince him of the unspeakable richness of the firmament.

A well-adjusted equatorial telescope will readily find anything in the following list from the position there given; otherwise recourse must be had to a good globe or map. The larger Star-Maps of the Society for the Diffusion of Useful Knowledge will answer the purpose excellently, as they contain all Smyth's objects, and are remarkable both for accuracy and cheapness : their great drawback is an unavoidable distortion towards the sides and corners, rendering those parts unlike the corresponding regions of the heavens, and requiring to be mentally corrected by comparison with the undistorted

central portion.* They contain all Flamsteed's stars, and the
7,646 of Piazzi's Catalogue,† with some others: the affixed
numerals are explained in the margin; but the student must
observe that the nº of any of Piazzi's stars must have the hour
of Right Ascension in which it occurs added to identify it, as
the same nº is repeated in each of the xxiv hours: to this
should be further affixed the letter P to designate the observer,
so that 142 P V will denote the 142nd star in Hour V of R A, in
Piazzi's Catalogue : the name of the constellation is often subjoined for the sake of convenience.   These maps also contain the
nebulæ of Messier and the two Herschels; and Smyth's character
of them, that they are " the best which have appeared," sufficiently marks their value.   In our little catalogue, the constellations being alphabetically arranged, precedence is given
to brightness in the larger star of each pair, and to general
conspicuousness in clusters and nebulæ, though great exact-

* An improved work on the plan of Jamieson's Celestial Atlas (1822)
is much wanted; every constellation of importance being separately
represented : the collection of materials from the heavens would be as
delightful an occupation for an amateur as their accurate arrangement
would be troublesome.   We have at this moment, as far as I know, no
complete popular representation of all the stars visible to the naked eye,
that is, down to 6 mag. inclusive, of which Humboldt says, there may be
from 5,000 to 5,800 (an extension to 9 mag. according to Argelander
would raise them to about 2,000,000).   Sir J. Lubbock admitted in 1836
that there were several stars, visible without the telescope, which do not
find a place in any of the standards of astronomy ; and Herschel II.
says that it is the case with every map he has ever used, that the leading
stars in the map are not those which catch the eye by their brightness in
the heavens.

† This catalogue, published at Palermo in 1814, includes the result
of nearly 150,000 observations made by Piazzi, requiring for their
reduction 30,000,000 figures.

ness in this respect has not been aimed at. Pairs whose connection is ascertained are termed "binary." The *magnitudes*
and *colours* of the Bedford Catalogue are everywhere adopted,
if it is not otherwise specified ; but comparisons are given from
the results of other observers ; and my own observations and
remarks are distinguished by brackets, at least where ambi-
guity might arise : some excuse may be alleged for their ad-
mission, since the eyes and instruments of those who may use
this treatise are much less likely to resemble Smyth's than
my own : observations also of little individual worth may
ultimately by accumulation and comparison acquire some
value. *Magnitudes* are very differently rated by different observers, and Struve has endeavoured to introduce a fresh scale,
including in 12 degrees the 20 of Herschel II. and Smyth :
those in the Bedford Catalogue are assumed from Piazzi for
the brighter star in each pair : where I have ventured to note
any discrepancy, it has been with a view to assist in detecting
variations of light : Schröter suggested, and Humboldt is of
his opinion, that variability may be the inseparable condition
of all light, and the evidence of its probability is continually
on the increase. As to estimates of *colour* there is also great
difference, arising from the differences of telescopes and eyes,
and even of the states of the same eye : still there are limits
of disagreement, and it is desirable to fix them, as there seems
reason to believe that these colours may change : where there is
any such suspicion, comparisons should be multiplied, and
their circumstances varied. As this is an interesting enquiry,
and one suited to amateurs, I had intended to insert many
more discrepancies between Smyth's colours and those of other
observers ; but I ultimately found that a very large proportion
may be reasonably referred to the causes just mentioned, and

included in the wide margin of those individual peculiarities of
perception or judgment which astronomers term "personal
equation," so that a few only have been retained, either where
there may be some suspicion of real alteration, or as specimens
of the differences to be expected in the enquiry. I have ventured to set aside, from their obvious peculiarity, Sestini's
colours, which caused a re-examination by Smyth, published
in his learned and elegant " Ædes Hartwellianæ," but I have
inserted some of Dembowski's, as their usual agreement gives
value to occasional discrepancies.* The colours of all the
objects in my list, as well as of many others, were carefully
compared with the Bedford Catalogue : my instrument and
experience were far inferior, and my eye usually biassed by
previous knowledge, so that I was little qualified for such
a scrutiny ; but with a great preponderance of agreement or
acquiescence, a few discrepancies were noted : dates are added,
as the idea of periodical changes of tint involves no impossibility. The *distances* between the stars (always from centre
to centre) are given from Smyth to fractions of seconds up to
10″, beyond that limit, to the nearest whole second.   Their
value to the student is the formation of an accurate eye.
The *angles of position*, which measure the inclination to the

* The Baron Dembowski, a very diligent observer of double stars at
Naples, uses a " dialyte " of 5½ feet focus and 5¼ inches aperture. This
is an ingenious modification of the achromatic, well known on the
continent, but neglected in England, though invented and put in practice here by Rogers, at the same time as by Littrow at Vienna, about 1828.
The object-glass is a single lens, and the correction for colour and figure is
made by a combination of two smaller lenses, placed at some distance
behind it : a construction which possesses some great advantages.
Plössl of Vienna has made many ; — one for the Sultan, 11½ inches aperture; and some of them are reported to be excellent.

meridian, of the line joining the stars, are omitted; the owners of micrometers, to whom they would be chiefly interesting, ought also to possess, in the Bedford Catalogue, better help than mine. The *range of visibility* depends of course on the light of the instrument and the sensitiveness of the eye. Our great observer Dawes has a sight capable of detecting very minute points with small optical means: with eyes and telescopes of average quality my experience leads me to state that the range of an achromatic of $3\frac{7}{10}$ inches aperture will terminate among 11 mag. stars (of Smyth's scale), though from some unknown cause,—perhaps, as Smyth suggests, peculiarity of hue,—smaller ones are sometimes to be caught. I have inserted some of these as tests. After Binary and Double Stars are placed the more conspicuous Clusters and Nebulæ : many of the great wonders, however, are beyond any common telescope.

The Right Ascensions and Declinations of the Bedford Catalogue have been brought up from 1840 to 1860, by allowing for the Precession of the Equinoxes, which by slowly carrying round the artificial network of meridians and parallels in front of the immoveable heavens, is continually changing the *nominal places* (not the *relative positions*) of the stars. The nearest minute of time and space was thought a sufficient degree of accuracy in this computation. The places of the additional objects have not been reduced; but the difference is not material. Globes and maps, admitting of no such reduction, will require allowance to be made on this account in proportion to their date: the Star Maps already recommended having been constructed for the same epoch as the Bedford Catalogue, will be found *behind* by quantities varying in different parts of the heavens, but nowhere much exceeding $\frac{1}{4}°$, and consequently never sufficient to cause mistake.

As to optical management : close pairs and crowded clusters gain by increasing the power ; so *in general* do dissimilar colours, and very minute points near larger stars; but experience will be the best guide. For difficult pairs we should follow Herschel I.'s advice, and adjust the focus previously upon a single star of nearly the same altitude, size, and colour; the peculiar aspect of the double star will be afterwards more striking. Occasionally a slight change of focus may relieve a weary eye. Large nebulæ always require low powers ; very small ones must be more magnified to shew their nature, and resolvable ones, to insulate their sparkling points. In most cases, low powers have the advantage from the beauty and variety of their broad fields. The magnifiers used by myself ranged from 64 to 250, in a few instances 300.

In the following list, under the head of Double Stars, the *Synonym*, which stands first, is either a Greek letter, which is Bayer's designation ; — an Arabic numeral, which is Flamsteed's ;— a compound title already explained in p. 154, which refers to Piazzi's Palermo Catalogue ; — or an Arabic numeral followed by Σ, the conventional symbol for the name of Wilhelm Struve, which refers to the great Dorpat Catalogue of Double Stars. In the class of Clusters and Nebulæ, an Arabic numeral, if followed by M, refers to Messier's Catalogue of Nebulæ in the "Connoissance des Temps" for 1783 and 1784; — if followed by ♅, the symbol of the name of Herschel I., to which a Roman numeral is added, it belongs to the corresponding catalogue of that observer ; — if by H alone, to the catalogue of his son. After the Synonym comes the *Place in the Heavens*, given first in hours and minutes of Right Ascen-

sion, then in degrees and minutes of Declination, marked N or
S, as the case may be; the Italic letters *n*, *s*, *p*, *f*, signifying
respectively north, south, preceding, and following, being
employed to indicate the relative positions of neighbouring
objects. Next are placed (in the case of Double Stars,) the
Magnitudes in corresponding Arabic numerals separated by a
comma; the half magnitudes being expressed by the decimal
point and figure ·5 (Struve's and Dembowski's magnitudes
are carried to minuter divisions). Last follow the *Colours*
in a similar order to that of the magnitudes.

All data, not specially excepted, are taken from the Bedford
Catalogue: for many descriptions of objects, and directions for
finding them, the author is responsible.

————✦————

ANDROMEDA.

This constellation is rich in interesting objects of every
class; it is however inconveniently high for an achromatic
telescope on the meridian, and should therefore be examined
some hours E. or W. of it, like many other similarly situated
regions.

Double Stars.

γ—1ʰ 55ᵐ   N 41° 39′—3·5, 5·5—11″— orange, emerald
green.   One of the most beautiful pairs in the heavens, though
stationary, and devoid of the interest of a binary system.   It
seems to have been first noticed by C. Mayer in 1778.   In
1842 Σ found the companion double, though so close that
common telescopes will not even elongate it, and the finest
only can divide it.   Sir W. K. Murray, with his 9 inch
achromatic, finds that these two minute stars differ in colour,

pale yellow, and smalt blue; and this is confirmed by Jacob
and Dawes; hence, by those who have seen it single, it has
been sometimes called blue, sometimes green.

π — 0^h 29^m N 32° 57'—4·5, 9—36"—white, blue.

56 and 203 P I—1^h 48^m N 36° 34'—6, 6—176"—both
yellow, 1834. Σ, p smaller and always deeper in colour, 1836.
[p larger and ruddier, 1850.] This pair, which forms a triangle
with an 8 mag. star, is a specimen of a class whose similarity,
notwithstanding their mutual distance, always impresses me
with the probability of some connection; more clearly deduced in the present case from a "proper motion" through
space, common to both.   Nearly pointed at by γ and β Triang.
at about twice their distance.

36—0^h 47^m N 22° 52'—6, 7—about 1"—bright orange,
yellow.   Binary.   Readily elongated by me, 1850, 1851, 1855,
shewing increased distance since Smyth's epoch, 1843.   Closely
$n p \eta$ towards ζ: visible to naked eye.

59—11^h 2^m N 38° 23'—6, 7·5—16"—bluish white, 1835
[yellowish white, 1850, 1855], pale violet. Σ egregie albæ,
1831.   I found not more than 1 mag. difference. Σ gives
6·7, 7·2, but to a different scale.

61 P II—11^h 14^m N 40° 46'—7, 11—50"—yellow, pale
lilac.   I missed the companion, but the object guides to a
pretty open 8 mag. pair, one of which is 62 P II.   Field fine
with low power.   About ⅓ from γ towards Algol.

175 and 176 P O—0^h 39^m N 30° 11'—8, 8—46"—both
pale yellow.   Curious similarity.   1½° $f \delta$.

22—0^h 3^m N 45° 18'—5—white, 1838 [clear yellow,1850].
A guide to an elegant pair nearly n,—8, 9—4·9"—pale yellow,
bluish.

240 P XXIII—xxiii^h 51^m N 23° 34'—8·5, 9—9·4"—pale

white, yellowish, 1833. [9 decidedly blue, with 1 mag. difference, 1850.] Σ gives 7·7,8·8, 1830.

[ψ or 20—about xxiii^h 38^m N 45° 30'—5,5—both white. Of
the same class with 56 and 203 P I.]

[A wide 7 mag. pair lies about half way from α towards ε.]

[About $\frac{1}{3}$ from γ Androm. towards β Pers. a little n of the
line, is a curious quadruple group of small stars, and $\frac{1}{2}$° n of
it, a double star.]

Nebulæ.

31 M—0^h 36^m N 40° 30'.   One of the grandest and least
resolvable in the heavens; a long oval, ill-defined, and brightening to the centre; so plain to the naked eye that it is strange
that the ancients scarcely mention it.   By moving the telescope rapidly to gain contrast Bond extends it to the surprising
dimensions of 4° in length and 2$\frac{1}{2}$° in breadth, of which common telescopes shew little, and less in proportion to the increase
of power.   Neither ♅, H, nor the Earl of Rosse with his 3
feet mirror could resolve it, though the latter perceived stellar
symptoms at its edge.   Bond's giant achromatic finds no resolution, though it is seen through a rich stratum containing above
1,500 stars.   He detects however two curious dark streaks,
like narrow clefts, both far beyond any ordinary instruments,
in which the darker of them forms imperceptibly the boundary
of the strongest light on one side : Secchi sees them both well :
Smyth notices an increase of brightness towards *sf* edge.   I
suspected more abrupt definition on that side, on which the
streaks lie ; but this was *after the knowledge of the fact*, which
has a great influence upon the eye ; the truth of ♅'s remark
being often exemplified, that a less degree of optical power
will *shew* an object afterwards, than would have been requisite
for its *discovery*.

M

32 M is in the same field with a low power; a small, hazy object, resolved into stars by the Earl of Rosse's 3 feet speculum.

18 ☿ V—0ʰ 33ᵐ N 40° 55′. Large faint oval nebula, requiring a low power: a large field (my 64 was barely large enough) includes it with 31 and 32 M.

ANSER.

This little modern asterism does not contribute much of general interest to the Bedford Catalogue; but the two following fields will give pleasure to all who seek out the glories of creation.

[6 and 8—about XIXʰ 22ᵐ N 24° 20′—4, 5—deep and pale yellow, in a beautiful field, 3° nearly s from β Cyg.]

[4, 5, 7 &c.— about XIXʰ 20ᵐ N 19° 30′. A magnificent region.]

ANTINOUS.

A constellation frequently mixed up with Aquila, containing some bright stars, of which η is variable, according to Argelander, in about 7ᵈ 4ʰ 14ᵐ; increasing in 57ʰ, but occupying 115ʰ in its decrease. Such inequalities are frequently met with among these mysterious objects.

Double Stars.

26 P XX—xxʰ 5ᵐ N 0ᵘ 27′—6·5, 7—3·5″—both white, 1832. Dembowski 7, 7·2,—white, green, 1853-4. [nearly equal, 1850: n a little larger with power 144.] 2ᵘ n of θ, a little f.

197 P XVIII—xviiiʰ 42ᵐ S 6° 4′—7, 9—99″—orange, cerulean blue. A smaller closer pair s p [white, blue]. Field beautiful with 80. 64 includes the cluster 11 M.

140 and 139 P XX—xxʰ 20ᵐ S 2° 34′—7·5, 8—60″—
both white, 1833 [white, grey, 1850, 1855]. Each has a
faint companion. Field, if large, very fine.

186 P XIX—xɪxʰ 29ᵐ S 10° 28′—7·5, 9—3·8″—pale
white, sky-blue. This requires favourable weather : it is the
largest of a group ƒ 37 Antin. In the field is another pair,
185 P XIX; triple, Smyth.

116 P XX — xxʰ 17ᵐ N 0° 37′—7·5, 12— 30″—white,
grey. Smyth says the companion is " so minute that its
distance is only an estimation." I saw it, however, steadily ;
so that it must be a good test : an isosceles triangle with
θ Antin. and 69 Aquil. the most s of two 7 mag. stars, near
together in finder.

12 P XX—xxʰ 4ᵐ S 0° 32′—8, 9—54″—both pale grey,
1835 [yellowish, bluish, 1850]. 1° n of θ.

274 and 275 P XVIII—xvɪɪɪʰ 56ᵐ S 0° 54′—9,9—26″—
both white. [s ƒ considerably the smaller, 1850.]

[139 and 140 PXX—about xxʰ 19ᵐ S 2° 40′. Three bright
stars, n with a very minute attendant, in a fine low-power field.]

[2654 Σ—xxʰ 6·7ᵐ S 4° 2′—6·2, 7·7, Σ 1826 [8, 9,
1855]—13·9″—both white [yellow ? blue ?]

[2661 Σ — xxʰ 10·4ᵐ S 2° 47′—about 8, 9—24″—
white, greyish.]

Cluster.

11 M—xvɪɪɪʰ 44ᵐ S 6° 26′. Fine cluster, like an expanded
fan, at the upper edge of the luminous cloud, which marks the
Shield of Sobieski. Smyth compares it to a flight of wild
ducks. An 8 mag. star is a little within its apex; an open
8 mag. pair s ƒ beyond it. Secchi, who fully resolves it, sees
it unaided in the Italian sky.

AQUARIUS.

A dull-looking constellation, but well repaying telescopic research.

Double Stars.

ζ—xxii^h 22^m S 0° 44′—4, 4·5 — 3·6″, 1831; 2·7″, 1842—flushed white, creamy. A very fine object; two suns revolving, Smyth thinks, in a period of 750 years, but as yet very uncertain. It occupies the centre of a triangle of nearly equal stars, all easily seen with the naked eye.

12—xx^h 57^m S 6° 22′— 5·5, 8·5 — 2·8″— creamy white, light blue. Brightest of the vicinity.

ψ¹—xxiii^h 9^m S 9ᵘ 51′—5·5, 9—49·5″—orange or topaz yellow, sky blue. Σ suspects common proper motion.

107—xxiii^h 39^m S 19 °27′—6, 7·5—5·5″—white, purplish. Possibly binary.

29—xxi^h 55^m S 17° 38′—6, 8, 1830—4·5″—both brilliant white. [very little differing in size, 8 perhaps the smaller, 1849, 1851, 1853, 1855.]

41—xxii^h 7^m S 21° 46′—6, 8·5—4·8″—topaz yellow, cerulean blue. A 7 mag. star makes it a pretty group.

94—xxiii^h 12^m S 14°13′—6,8·5—14″—pale rose, light emerald, 1838; orange, flushed blue, 1850. Σ ascribes common proper motion to this beautiful pair.

53 — xxii^h 19^m S 17° 27′— 6·5, 6·5 — 9·9″— both pale white.

200 P XXII— xxii^h 36^m S 9° 3′— 7, 8·5 — 2·7″— both white, 1833, 1838 [8·5 grey or bluish? 1850]. 2½° p λ, a little s.

219 P XXII. Triple— xxii^h 41^m S 4° 57′—7·5, 8, 9—4·2″, 55″—yellow, two flushed white.

[AQUARIUS]

[τ²—about xxıı^h 41^m S 14° 30′ — is a beautiful orange 5 mag. star, with a distant companion.]

[24—ƒ 2 M, a little *n*, in the head—7, 10.]

[About half way between β ♒ and ε Peg. lies a pretty little white 8·5 mag. pair.]

[About xxı^h 38^m S 0° 30′ a rich region will be found, where a low power will include three double stars at once.]

Nebulæ.

2 M—xxı^h 26^m S 1° 27′. Beautiful large round nebula, shewing, with $3\frac{7}{10}$ inches aperture, a granulated appearance, the precursor of resolution. H compares it to a heap of fine sand, and considers it to be composed of thousands of 15 mag. stars. Smyth observes that " this magnificent ball of stars condenses to the centre, and presents so fine a spherical form, that imagination cannot but picture the inconceivable brilliance of their visible heavens, to its animated myriads."

1 ♅ IV—xx^h 57^m S 11° 55′. Planetary; somewhat elliptic; very bright for an object of this nature; pale blue; not well defined in $5\frac{1}{2}$ feet achromatic, but bearing magnifying like a planet, much otherwise than a common nebula. One of the finest specimens of these extraordinary bodies, which their discoverer ♅ removed into a distinct class. The Earl of Rosse finds a ray on each side. Secchi, who gives its diameters 25″ and 17″, sees it sparkle, and believes it to be a mass of stars: this he thinks also indicated by the bluish tint common in planetary nebulæ, which is usually met with in groups of very minute stars. Can we view this as a confirmation of Olbers's ingenious idea, countenanced by Σ and Steinheil, of the imperfect transparency of space, in which blue light may possibly be least subject to absorption ?

AQUILA.

Altair, the *lucida* of this rich constellation, has been thought variable, and has a very sensible proper motion.

Double Stars.

γ—xix^h 40^m N 10° 16′—3—pale orange. Smyth gives a 12 mag. attendant, which I did not make out, but have inserted the object for the sake of its beautiful field. It is now brighter than β, which implies a change in one of the stars; though in many instances Bayer, who affixed the Greek letters in 1603, seems not to have been entirely influenced by magnitude.—δ is also in a beautiful neighbourhood.

π—xix^h 42^m N 11° 28′—6, 7—1·7″—pale white, greenish, 1831, 1836. Σ subflavæ, 1829. Dembowski both white, 1856. I found this pretty pair was not single with 80, very close with 144 : a good test.

15—xviii^h 58^m S 4° 14′—6, 7·5—35″—white or yellowish white, red lilac. 1° nearly *n* from λ Antin.

23—xix^h 11^m N 0° 50′—6, 10—3·1″—light orange, grey. ♓ and Smyth notice the increasing visibility of 10 with higher powers, which struck me independently. I had not perceived it with 80 ; it was distinctly seen with 144. Possibly there may be some peculiar quality in its light : it is an elegant object, but requiring fine weather; easily found from δ and ν.

56—xix^h 47^m S 8° 56′. *n f* from this star lies an open 6 mag. pair. Smyth calls them both blue, 1834: the smaller appeared so to me, 1855, but the *n*, or larger, evidently of a different tint, perhaps pale greenish yellow.

57—xix^h 47^m S 8° 35′—6·5, 7—35″—both blue, 1834. Σ egregie albæ, 1833. Distinctly contrasted to me, 1851 ; pale yellow, pale lilac, " colours entirely different," 1855 ; —

a totally independent observation, as I had not previously identified the object; this pair should be watched, as two of the first observers have attested the similarity of the colours.

5. Triple—xviii^h 39^m S 1^v 6'—7, 8, 14—13", 30"—white, lilac, blue.   14 is so minute, Smyth says, as to have escaped former observers except ♅. P, however, saw it with far inferior means; and I have on several occasions, from 1850 to 1852, seen it more or less distinctly by averted vision, though 14 mag. was usually far beyond my reach: it rather improves with magnifying.   5 is η Serp. of some maps.

11—xviii^h 53^m N 13^v 26'—7, 10— 19"—pale white, smalt blue.   Smyth thought 7 might have been rated 6.   It lies equilaterally with ε and ζ.

2 P XX — xx^h 3^m N 16° 30'—7, 10—5·9"—pale topaz, lucid blue.   ♅, red, deeper red; but he was partial to red tints.   I thought 7 white.   Perhaps binary.   Close to Sagitta.

144 P XIX — xix^h 23^m N 2° 37'—7, 11, 1838—37"— deep yellow, pale green. [11 readily visible, 1850, 1855, even considerably out of focus, and several hours past meridian, as if 10 mag.; the colour also, previously known, was evident.] 1° *f* δ, furthest of two *f* that star.   I found a bright and pretty pair in the same field with a low power.

302 P XVIII—xviii^h 59^m N 6° 20'—7·5, 9—10"—lucid white, cerulean blue.   Brightest in the vicinity.

241 P XIX — xix^h 36^m N 8° 3'—7·5, 9·5, 1833— 27"— pale topaz, lilac. [small star larger than 9·5 ? 1850. P rated it 10.   Equilateral with α and γ.   1° *p* lies a pretty pair— 9, 10: and again, not far *s p*, a very fine field.]

257 P XIX — xix^h 38^m N 10° 26'— 8, 10 —4"—white, smalt blue.   Closely *n p* γ.

43 and 44 P XX—xx^h 7^m N 6° 10'—8·5, 8·5, 1833—44"

[AQUILA]

both lucid white. [*s* star smallest, 1850; the reverse of Smyth.]

263 P XVIII — xviii^h 54^m   N 14° 43′—8·5, 10·5—6·5″— pale yellow, 1836 [white, 1850], sapphire blue.   Smyth calls this a handsome test-object. I found 10·5 occasionally distinct, though long past meridian.   Closely *f* ε, 3 mag., a beautiful yellow star; but far *s* in a fine field.

[About ⅓ from ζ Aquil. towards α and β Sagitt. are two 6 mag. stars, near together, visible to naked eye: the *n p* of these is a fine wide pair—6, 8 — yellow, pale lilac.]

[15—6 mag.—yellow—1° *n* of λ Antin. forms a grand pair with a reddish purple star.]

[68—about xx^h 20^m   S 4° 50′—7 mag. is one of a pretty triangle : a little. *s p* 69, 5 mag.]

[About xx^h 28^m   N 1° lies a triangle, the most *s* of which is a wide pair, — 7, 9 or 10 — orange, blue. All the Galaxy in Aquila is strewed with pairs and groups of stars.]

Cluster.

2024 H—xviii^h 49^m   N 10° 11′. An interesting field.

ARGO NAVIS.

Not a remarkable constellation, excepting for the crowded part of the Galaxy which it includes.

Double Stars.

149 and 147 P VII — vii^h 28^m   S 23° 10′— 6, 6—10″— both topaz-tinted. [*n p* rather larger, 1851, 1856; otherwise than Smyth, 1831.]

72 and 74 P VIII—viii^h 19^m   S 23° 36′—6, 9·5—45″— orange, bluish green.

175 and 177 P VII—vii^h 33^m   S 26° 29′—6·5, 6·5—9·8″

—both topaz yellow. [both stars much larger, 1851, than last pair but one, as if the mags. had been accidentally reversed.]

2—VII^h 39^m S 14° 21'—7, 7·5—17"—silvery white, pale white. In field with 4, 5 mag., pale yellow.

5—VII^h 41^m S 11° 51'—7·5, 9—3·5"—pale yellow, light blue, 1834. Σ 9 cærulea, 1831. [9 ruddy? 1851.]

Clusters and Nebulæ.

38 ♀ VIII—VII^h 30^m S 14° 11'. Grand broad group, visible to naked eye, too large even for 64 : some brilliant 5 or 6 mag. stars, and neat pair—7·5, 8—8"—both bright bluish white. About 2½° p a group round 4.

46 M—VII^h 35^m S 14° 30'. Beautiful circular cloud of small stars about ½° in diameter : best with low power : a little p the group round 4, nearer to it than 38 ♀ VIII. I could not see a planetary nebula which it contains, and which in Lassell's 20 feet reflector is "an astonishing and interesting object."

93 M—VII^h 39^m S 23° 32'. Bright cluster in a rich neighbourhood.

11 ♀ VII—VIII^h 4^m S 12° 27'. Large loose cluster of stars, chiefly about 10 mag., closely n p 19 Arg., a 6 mag. yellow [bright orange] star, attended by a fine group. 19 seems larger than 6 mag. to my unaided eye.

37 ♀ VI—VII^h 53^m S 10° 24'. Fine broad starry cloud, from 10 mag. down to mere nebulosity ; much better with 64 than higher powers. Vicinity gorgeous.

64 ♀ IV—VII^h 36^m S 17° 53'. Bright planetary nebula ; pale bluish white ; 12" or 15" in diameter to ♀. With my 64, like a dull 8 mag. star : with more power, small, brilliant, undefined, surrounded with a little very faint haziness. In a glorious neighbourhood.

ARIES.

Three stars near together mark it to the naked eye, but it reaches some way further E. into a dull region.

Double Stars.

γ—1ʰ 46ᵐ   N 18° 36′—4·5, 5—8·8″—bright white, pale grey or faint blue.   Discovered by Hooke in 1664, in following the comet of that year.   " I took notice that it consisted of two small stars very near together ; a like instance to which I have not else met with in all the heaven."   A good object for small telescopes.

λ—1ʰ 50ᵐ   N 22° 55′—5·5, 8—37″—yellowish white, blue. Pointed at by γ and β.

30 and 128 P II — 11ʰ 29ᵐ   N 24° 2′—6, 7—38″—topaz yellow, pale grey.   " The most southern of a group of about a dozen double stars, spread over the adjoining portions of the three constellations, Aries, Musca, and Triangulum, with extensive patches of dark and blank space between them." (Smyth.)

179 P I—1ʰ 42ᵐ   N 21° 35′—6, 8′—2·4″—topaz yellow, smalt blue.   The yellow, I think, has a scarlet glare around it, like Hind's New Star in Oph. when brightest.   Dembowski white, 1854.   80 just divides this beautiful pair.   2ᵛ n p β.

33 — 11ʰ 32ᵐ   N 26° 27′—6, 9—28·5″—pale topaz, light blue.   9 seems very small.   In *Musca*.

10—1ʰ 56ᵐ   N 25° 16′—6·5, 8·5—2·2″—yellow, pale grey. Double, 250 ? if so, very difficult.   Between α ♈ and α Triang. nearest former star.

AURIGA.

The leader of this beautiful constellation, Capella, is very brilliant.   H and Σ think it has increased.   H classed it decidedly above Wega in 1847, otherwise than he had formerly,

and therefore second in the heavens. So Galle and Heis. With
me, Wega takes precedence; but the objects are distant, and
differ in colour, white and sapphire, and, as Smyth observes,
this difference may influence estimates of size. Ptolemy,
El Fergani (10th cent.), and Riccioli, have all called Capella
red. Its parallax is almost imperceptible, proving amazing
distance, far exceeding that of some smaller stars. Such instances shew the insecure foundation of ♓'s theory as to the
regular distribution of stars in space. See 37 M *infra*.

Double Stars.

14—V^h 6^m N 32° 31'—5, 7·5—14"—pale yellow, orange,
1832; greenish yellow, bluish yellow, 1850. Σ subviridis,
albacærulea, 1830. [pale yellow, lilac, 1850.

26—V^h 30^m N 30° 24'— 5, 8 — 12"— pale white, violet.
3° *n f β* ♉.

ω — IV^h 50^m N 37° 41'—5, 9 —7"—flushed white, light
blue, 1831, 1833, 1850. Σ subviridis, albacærulea, 1828.
[white, ruddy, 1850; an unusual combination, plainer with 80
than 250.]

41—VI^h 1^m N 48° 44'—7, 7·5—8·2"—silvery white, pale
violet. Relatively fixed, common proper motion : a remarkable though not infrequent phænomenon.

61 ♓ VIII—IV^h 59^m N 37° 10'. Cluster, *p* a neat pair, 644 Σ
—7, 8— 1·8"—topaz, amethyst. 80, in contact; 250, divided.

225 P V—V^h 41^m N 31° 50'— 8, 8·5 — 3·8"— creamy
white, pale grey.

[941 Σ—VI^h 26^m N 41^υ 43' according to H and South,
1824—9, 10 [I thought 8, 9·5]—1·7"—Σ alba cærulea, alba
subpurpurea. A good test; the most *n* of two stars in finder,
1^υ *s* of 50 Aur., 5 mag.]

[φ or 24, 5 mag., lies 6° *n* of β ♉, in a superb region.]

[AURIGA]

Clusters.

37 M—v^h 43^m N 32° 35'. Smyth calls this "a magnificent object : the whole field being strewed as it were with
sparkling gold-dust; and the group is resolvable into about
500 stars, from 10 to 14 mags., besides the outliers." Even
in smaller instruments extremely beautiful, one of the finest of
its class. All the stars in this globe must be nearly at the
same distance from us, and consequently their real sizes must be
different. The aspect of the Nubecula Major in the S. hemisphere convinced H of this : it is ocular proof of the fact—
deducible from parallax and common proper motion, yet not
sufficiently attended to — that, among the stars, apparent
magnitude, and distance from the Earth, are quite unconnected.

38 M—v^h 20^m N 35^u 46'. Noble cluster, arranged in the
form of an oblique cross; a pair of larger stars in each arm,
and a bright single star in centre. [A small round cluster of
minute stars lies *s p* in the same low-power field, just *f* a
coarse 7 or 8 mag. pair, which again is *s f* a 6 mag. star,
orange, in a glorious field.]

36 M—v^h 27^m N 34° 3'. Beautiful assemblage of stars,
very regularly arranged. 2° *f* φ. Wrongly numbered 56 in
Map of Society for Diffusion of Useful Knowledge.

39 ♅ VII—v^h 19^m N 35° 12'. In a splendid district.
A little *n f* is a superb low-power field.

33 ♅ VII — v^h 10^m N 39° 12'. A magnificent region.

BOÖTES.

A fine constellation, of which the leader, Arcturus, is rated
next to Sirius, and before Wega, by ♅ and H. A noble object
at all times, but never so interesting as when, enveloped in the

tail of Donati's comet, 1858, Oct. 5, and only 20' from the nucleus, it flashed out so vividly its superiority. Smyth calls it reddish yellow; it is golden yellow to me: Schmidt thinks it has of late years lost all redness, and is growing paler. The first star seen in daytime, by Morin, 1635. It has a great proper motion of more than 1" R A, and nearly 2" Dec. annually, so that, as Humboldt says, it has moved 2½ times the Moon's diameter since the days of Hipparchus. Yet its parallax is insensible. How inconceivable, then, must be its dimensions and its speed!

Boötes is rich in pairs, poor in clusters and nebulæ.

Double Stars.

ε—xiv^h 39^m N 27° 40'—3, 7—2·9"— pale orange, seagreen. This "lovely object," as Smyth calls it, is probably binary of long period : a well-known test for moderate telescopes. I have seen it perfectly with 28 inches focus, 2¼ inches aperture, by Bardou, of Paris. 80 of the 5½ feet achromatic was sufficient to divide it in fine weather.

ζ—xiv^h 34^m N 14° 20'— 3·5, 4·5—1·3"— bright white, bluish white. Excellent test: with 5½ feet achromatic, 144 and 250, elongated, or discs in contact.

π—xiv^h 34^m N 17° 1'—3·5, 6—6"—both white, 1836, white and creamy, 1850: several observers see a slight difference.

ξ—xiv^h 45^m N 19° 41'—3·5, 6·5—7"—orange, purple, 1831–1842. Dembowski yellow, orange, 1856. Binary: period 117 years, H.

δ—xv^h 10^m N 33° 50'—3·5, 8·5— 110"—pale yellow, light blue.

μ^1—xv^h 19^m N 37° 52'— 4, 8 — 109"— flushed white,

greenish white. The companion, μ^2,—8, 8·5—is very closely double ; binary, of doubtful period : probably all form one system, a Sun, Earth, and Moon of unborrowed light.

ι—xiv^h 11′ N 52° 1′—4·5, 8—38″—pale yellow, creamy white. Σ thinks they have common proper motion. 4·5 is again double in the Czar's telescope at Poulkova.

44—xiv^h 59^m N 48° 12′— 5, 6 — 3·7″, 1842 — pale white, lucid grey, 1842 ; yellow, cerulean blue, 1850. Σ sub-flava, subcærulea, 1832. Fletcher white, yellow, 1851. Miller both white, 1853. Dembowski yellow, orange, 1856. [yellow, ruddy or purplish, 1850.] Σ and Argelander have found variable light here. Certainly binary ; with unknown period, and widening. [more distant than the following, 1850.]

39—xiv^h 45^m N 49° 18′— 5·5, 6·5 — 3·8″—white, lilac. [scarcely 1 mag. difference, 1850.] A little $n\,p$ 44.

κ—xiv^h 8^m 27^s N 52° 27′— 5·5, 8 — 13″— pale white, bluish.

69 P XIV—xiv^h 17^m N 9° 5′— 6, 7·5 — 6·3″—flushed white, smalt blue, 1835. Dembowski golden yellow, rose tint, 1853, 1855. [6, pale yellow ; 7·5, sometimes blue, more usually tawny, 1854. This uncertainty of hue, which I have found troublesome in the smaller components of some pairs, may probably arise from contrast with the tinted background given by the " outstanding " blue rays in the light of the larger star, combined with unequal sensitiveness to colour in different states of the eye, which may sometimes be most im-pressed with the real, sometimes with the complementary hue.]

1 — xiii^h 34^m N 20° 40′— 6, 10—4·9″— sapphire blue, smalt blue. 161 P XIII, 7 mag., bluish, in field.

279 P XIV—xv^h 1^m N 9° 46′—7·5, 7·5—4″—both pale white.

[BOÖTES]

220 and 219 P XIII—xiii^h 44^m N 21° 58′—7·5, 8—86″— both flushed white, 1831. I saw some difference in colour, 1852; perhaps yellowish, and bluish white: but very little in magnitude, each about 7; 220 rather the larger. In the same field with 6, a fine yellow 5 mag. star, marked only 6 mag. on Map of S.D.U.K., but too low, being very visible to naked eye.

[About xv^h N 54° is a fine open pair—pale blue? grey? just visible to naked eye.]

CAMELOPARDUS.

Widely extended, but obscure; containing a few good objects.

Double Stars.

269 P IV—v^h o^m N 79° 4′—5·5, 9—34″—light yellow, pale blue. Perhaps binary.

232 and 230 P XII—xii^h 48^m N 84° 10′—6, 6·5—22″ —both bright white, 1833. Σ egregie albæ, 1832. [some difference, 1852, pale yellow? pale violet?]

1—iv^h 21^m N 53° 36′—7·5, 8·5 — 10″—white, sapphire, 1838: so Dawes. Dembowski green or blue, pale rose, 1854. Nearly half-way from α Pers. to δ Aur., a little n of line; in an ill-marked part of the sky.

159 P VII—vii^h 33^m N 65° 29′—8, 8—16″—both white.

[97 P III—about iii^h 30^m N 59° 30′—6, 9—orange with scarlet glare, blue.]

Nebula.

53 ♅ IV— iii^h 55^m N 60° 27′. Planetary: small and dim. Closely f, ½° n, is 47 ♅ VII, an elegant group with 7 mag. pair, preceded by another, wider apart.

CANCER.

A constellation marked only to the naked eye by the re-markable cluster Præsepe.

Double Stars.

σ^2—VIII^h 46^m N 31° 6'— 5·5, 7 — 1·4"— white, yellow. Excellent test: elongated with 80, divided with 144 of my 5½ feet achromatic. ι^2 of Flamsteed, ᛏ, H, South, Σ, who finds difference of magnitude variable.

ι—VIII^h 38^m N 29° 16'—5·5, 8—30"—pale orange, clear blue. Beautiful contrast.

ζ. Triple — VIII^h 4^m N 18° 4'— 6, 7; 7·5 — 1·2", 5"— yellow, orange, yellowish, 1832–1843. Dembowski all white, 1854-5-6. Binary; close pair revolving in about 60 years, companion in about 500 or 600 years. I elongated the former with 80 and 144, and saw discs in contact with 250, 1849. It seems to have become considerably more difficult now, 1859. An excellent test; pointed at by α and β II, at twice their distance.

φ^2—VIII^h 18^m N 27° 23'—6, 6·5, 1833, 1843—4·8"—both silvery white. Dembowski both yellow, 1854. [much less unequal, 1849; nearly equal, 1856; still more striking if compared with v^1.] Σ gave 6, 6·5 of his scale, 1826. Dembowski both 6·5, 1854.

σ^4—VIII^h 53^m N 32° 48'— 6, 9 —4·8"—lucid white, sky-blue. 66 of Map?

v^1—VIII^h 18^m N 25° 0'—7, 7·5—5·8"—pale white, greyish. Σ gives 1 mag. difference; Dembowski almost as much : they seem considerably unequal.

1177 Σ—according to him, VII^h 54·9^m N 28° 3'—7, 8 — 3·8"—no colour in Smyth [white, greyish⁷.

[194, or 1311 Σ—in his work, VIII^h 57·6^m N 23° 38'—6·7, 7·1—7·2"—both white.   A little n p ξ, 5 mag. not in Map.]

[1228 Σ—VIII^h 17·3^m N 28° 6'—8, 8·5—8·93"—both white.   Not far from line joining ϕ^1 and ϕ^2, nearer the former star.]

Cluster.

44 M—VIII^h 30^m N 29° 15'.   The Præsepe of the ancients, just resolved by the naked eye; too large for ordinary powers, but full of fine combinations: two triangles will be remarked. Galileo counted 36 stars in it, with his newly-constructed telescope.

67 M—VIII^h 44^m N 12° 19'.   Cluster of stars of various magnitudes; ♓ counted more than 200.   Visible in finder.

CANES VENATICI.

The nebulæ here are fine.   The only prominent star comes first on the list.

Double Stars.

12 (Cor Caroli) — XII^h 49^m N 39° 4'— 2·5, 6·5 — 20"— flushed white, pale lilac, 1837; full white, very pale, 1850. Σ albæ, 1830.   H, "with all attention I could perceive no contrast of colours," 1830–1.   [white or yellowish, tawny or lilac, 1850.]   Relatively fixed for 57 years, yet considerable proper motion.   Here again are unequal stars which must be at the same distance from us.

2 — XII^h 9^m N 41° 26'— 6, 9 — 11"— golden yellow, smalt blue [6, orange with scarlet glare, a peculiar hue sometimes found].   A striking though not conspicuous object, $\frac{1}{3}$ from Cor Caroli towards δ Ω.

Nebulæ.

51 M — xiii^h 24^m N 47° 55'. Earl of Rosse's wonderful spiral : its wreaths of stars are of course beyond all but the first telescopes ; common ones will only shew two very un-equal nebulæ nearly in contact, both brightening in the centre : traces of the halo encompassing the larger may perhaps be caught.   They form a misty spot in finder, 3° *s p* Alkaid, at the end of the Great Bear's tail.

3 M—xiii^h 36^m N 29° 4'.   " A brilliant and beautiful globular congregation of not less than 1,000 small stars " (Smyth), *blazing* splendidly (that is running up into a con-fused brilliancy) towards the centre, with many outliers.   I could hardly resolve it.   It stands in a triangle of stars, rather nearer Arcturus than Cor Caroli.

94 M—xii^h 44^m N 41° 53'.   Small bright nebula like a comet ; resolvable, Smyth : a little *p* Cor Caroli, 2½° *n.*

63 M—xiii^h 10^m N 42° 46'. Oval, not bright. ♃'s tele-scope shewed it 9' or 10' long, and near 4' broad, with a very brilliant nucleus.   An 8 mag. star is grouped beautifully with it.

CANIS MAJOR.

α (Sirius) is the leader of the host of heaven : a glorious object, in all likelihood either far greater or more splendid than our Sun.   Its colour has probably changed.   Seneca called it redder than Mars ; Ptolemy classed it with the ruddy Antares (α ♏).   I now see it of an intense white, with a sapphire tinge, and an occasional flash of red.   Hind and Pogson have found similar decided changes of colour in variable stars. From irregularities in the proper motions of Sirius and Procyon (*a* Can. Min.) Bessel fully believed that each was a member of a binary system, their companions being invisible dark bodies: and subsequent observations seem to confirm the wonderful idea. The minuteness of the parallax of Sirius indicates an inconceivable distance, far greater, probably, than that of many fainter stars.

Double Stars.

μ—VI^h 50^m S 13° 52′—5·5, 9·5—3·5″—topaz yellow, grey. 3° *n f a*, among rich fields.

ν¹—VI^h 30^m S 18° 33′ — 6·5, 8 — 17″ — pale garnet, grey. 3° *s p a*.

30—VII^h 13^m S 24° 42′—6·5, 9—85″—white, pale grey. In a rich cluster, 17 ♅ VII. [About 1° 50′ *n*, 1° *p*, lies a very fine pair—6·5, 8—fiery red, greenish blue.]

Clusters.

41 M — VI^h 41^m S 20° 36′. Superb group, visible to naked eye, 4° beneath *a*.

12 ♅ VII—VII^h 11^m S 15° 23′. Beautiful cluster melting into a very rich neighbourhood, as though the Galaxy were here approaching us. 64 includes a bright white star *p*. Smyth observes that the stars are nearly all 10 mag. 3° *f γ*.

CANIS MINOR.

a (Procyon) a fine pale yellow star (which see mentioned under Sirius), has a curious variable star about 2½′ distant, very nearly *f*.—8 mag. Smyth, 1833. 9 mag. Fletcher, 1850. Missed by Bond, 1848, Fletcher, 1853, Hind, 1853-4-5. There are other minute stars in the field.

Double Stars.

14. Triple—vɪɪ^h 51^m   N 2° 36′—6, 8, 9—75″, 115″—
pale white, 1831, bluish, blue [6 deep yellow, 1851, and 8
very little brighter than 9].

170 P VII—vɪɪ^h 33^m   N 5° 33′—7, 8—1·4″—white, ash-
coloured. 144 shewed discs in contact; 250 now and then
just split them : an admirable test, lying most conveniently *f*
a little *s* of *a*, and being the largest star in that direction. It
is often called 31 Can. Min. Bode. Possibly binary.

CAPRICORNUS.

Not a conspicuous constellation, but containing some good
objects, among which its principal star takes a high rank.

Double Stars.

*a*² and *a*¹—xx^h 10^m S 12° 59′—3 (*a*²), 4 (*a*¹)—373″—pale
yellow, yellow. A noble pair, easily distinguished without
a telescope.

β ²—xx^h 13^m S 15° 13′—3·5, 7—205″—orange-yellow,
sky-blue.

ρ—xx^h 21^m S 18° 16′—5, 9—3·8″—white, pale blue, 1830;
a 7·5 mag. yellow star follows considerably *s*. [pale yellow,
ruddy purple, 1850 : 7·5, which makes a fine addition, seems
violet or lilac.] Secchi finds movement here.

σ—xx^h 11^m S 19° 33′— 5·5, 10—54″ — yellow, violet,
1837. [orange, blue, 1850 : the companion underrated ?]

o²—xx^h 22^m S 19° 3′—6, 7—22″—both bluish, 1832
[white, bluish, 1850].

[CAPRICORNUS]

[About xx^h 36^m S 16° is a pretty pair—8, 9 — lilac, perhaps bluish green. 1¼° f 240 P XX, a little n.]

Nebula.

30 M—xxi^h 32^m S 23° 47'. Moderately bright : beautifully contrusted with an 8 mag. star beside it : like a comet with 64, with higher powers resolvable. A brilliant cluster to ♅. " What an immensity of space is indicated ! Can such an arrangement be intended, as a bungling spouter of the hour insists, for a mere appendage to the speck of a world on which we dwell, to soften the darkness of its petty midnight? This is impeaching the intelligence of Infinite Wisdom and Power, in adapting such grand means to so disproportionate an end. No imagination can fill up the picture of which the visual organs afford the dim outline; and he who confidently probes the Eternal Designs cannot be many removes from lunacy." (Smyth.) It lies closely p, a little n, from 41, a 5 mag. star.

CASSIOPEA.

Here lie a multitude of very rich Galaxy fields. The leader a is slightly variable. Snow always found it sharper and smaller, and more readily obscured by fog, than β or γ, even when equally bright.*

Double Stars.

γ—0^h 48^m N 59° 57'—3 mag. Inserted for its beautiful contrast with the minute surrounding stars.

η—0^h 41^m N 57° 4'—4, 7·5—9·1"—pale white, purple,

---

* A difference in the *aspect* of different stars, independent of magnitude, and sometimes of colour, has been noticed by several observers. Smyth, speaking of a dull 11 mag. star seen best by averted vision,

1843. H and South red, green. Σ flava, purpurea, 1832. Fletcher yellow, purple, 1851. Miller white, dusky orange, 1851. [yellow, garnet, 1850.] Binary : period about 700 years.

72 P II. Triple—11ʰ 18ᵐ N 66° 46′—4·5, 7, 9—2·1″, 7·5″—pale yellow, lilac, fine blue.   55 of ♅, ι of others. Fine but not easy object with 5½ feet achromatic.

ψ—1ʰ 16ᵐ N 67° 24′—4·5, 9—32″—orange, blue.   9 again double in large telescopes.

101 and 100 P XXIII—xxiiiʰ 24ᵐ N 57° 47′—5, 7·5—74″ —light yellow, white, 1830 [7·5 pale lilac, 1854]. Dawes has doubled it again. [A 10 mag. pair ƒ.]

σ—xxiiiʰ 52ᵐ N 54° 58′—6, 8—3″—flushed white, smalt blue. In a glorious low-power field.

κ—about 0ʰ 24ᵐ N 62°—4—bright yellow, 1837 [white, 1850; a little yellowish, 1855] is a guide to a grand vicinity [one group resembles the letter Y].

φ—about 1ʰ 10ᵐ N 57° 30′—5—is attended by a beautiful group, 42 ♅ VII.

55—11ʰ 4ᵐ N 65° 52′—6—is s of the spot where the Great New Star flamed out, Nov. 1572, speedily rivalling Venus, so as to be seen at noon-day, then fading during 16 months to utter extinction : there is some idea that similar appearances took place here in 945 and 1264; if so, 1872 may

remarks that "there are many of much smaller magnitudes which shine quite sharply, and emit a strong blue ray." Argelander gives ζ Aurig. as a striking instance, among others, of peculiarly intense light for its magnitude. He says that red and yellow stars appear brighter in proportion to the superiority of the eye and instrument. Schmidt finds that red stars gain in twilight, lose by night, as compared with white ones, and that the position of the eyes is of material consequence in such estimates.

possibly witness a repetition of this incomprehensible phæno-menon. Its colour changed from white through yellow and red to blue. Hind thinks several variable stars increase blue, are yellow after maximum, and flash red in decreasing.

Clusters.

31 ♅ VI—1ʰ 37ᵐ   N 60° 32′.   Visible in finder ; a very good field with 64 ; 80 shewed a little pair, measured by Smyth. 2½° from δ, on a line drawn from α through δ.

103 M—1ʰ 24ᵐ   N 59° 58′.   Beautiful field.   1° *f* a little *n* of δ.   Omitted in Map of S. D. U. K.

78 ♅ VIII—0ʰ 35ᵐ   N 61ᵛ 1′.   Fine cluster, somewhat like the letter W ; half way from γ to κ.

30 ♅ VI—xxiiiʰ 50ᵐ   N 55° 56′.   Beautiful large faint cloud of minute stars : " a mere condensed patch " as Smyth remarks, " in a vast region of inexpressible splendour, spread-ing over many fields " ; including the whole Galaxy through this and the adjacent constellations.   [A beautiful group in a rich field lies about ¾° *f* δ.]

CEPHEUS.

Much more barren to the naked eye than to the telescope.

Double Stars.

β—xxiʰ 27ᵐ   N 69° 57′—3, 8—14″ — white, blue, 1833, 1843.   Dembowski yellow, violet, 1852, 1854. [white, blue, 1850.]

δ—xxiiʰ 24ᵐ   N 57ᵛ 42′—4·5, 7—41″—orange or deep yellow, fine blue.   An especially fine pair, somewhat like β Cyg. 4·5 is slightly variable : period 5ᵈ 8ʰ 30ᵐ : Argelander adds 18ᵐ.   Schmidt suspects variation in many stars of Cepheus.

[CEPHEUS]

κ—xx^{h} 14^{m} N 77° 17′—4·5, 8·5—7·5″—bright white, smalt blue.*

ξ—xxii^{h} o^{m} N 63^{v} 57′—5, 7—5·8″—both bluish, 1839; flushed, pale lilac, 1851. Σ subflava, cærulea, 1831. Dembowski white, violet, 1854. [white, tawny or ruddy, 1850.] About 1° *n*, and as much *p*, is 2843 Σ, in his work 7·2, 7·5—2·2″—albæ ; this I have not seen, but it must be delicate and beautiful.

11 and 12 P XXII—xxii^{h} 4^{m} N 58° 37′— 6, 6·5, 1839 [very nearly equal, 1850]—21″—both white.   Σ finds 6·5 again double. Just *p* λ : a little *n* of ζ.

248 P XXI. Triple—xxi^{h} 35^{m} N 56° 51′—6, 8·5, 8·5—12″, 20″—pale yellow, two grey.

191 P II—ii^{h} 48^{m} N 78° 52′—6, 10·5—5·2″—orange, smalt blue. Good test for a moderate instrument.

o—xxiii^{h} 13^{m} N 67° 21′—7, 9—2·5″—orange yellow, deep blue.

285 P XXI—xxi^{h} 39^{m} N 58° 8′. The celebrated "Garnet sidus" of P, visible to naked eye, but variable, and hence probably omitted by Flamsteed. ♄ says "it is of a very fine deep garnet colour," especially after viewing a white star, such as α. 2½° *s* of ν.

* A peculiarly strange coincidence is too remarkable to be omitted, attested by an unimpeachable witness, the late Mr. Baily. "The RA of this star was erroneously calculated by the two separate and independent computers, who *agreed precisely in every figure:* and the error was even unobserved by the vigilant eye of Mr. Stratford. On recomputing the place of this star, with a view to discover the cause of its discordancy, I myself fell into exactly the same error, and obtained *precisely the same figures;* and it was only on going over the operation a second time that I *accidentally* discovered that we had all inadvertently committed the *same* mistake!"—Memoirs of R. Astron. Soc. IV. 290.

CETUS.

The largest, but far from the most interesting of the constellations. Its alphabetical leader a is not now so bright as β: probably one or both have changed. a is worth looking at, as a fine combination of a beautiful 2·5 mag. orange star [yellow, 1856] with a 5·5 mag.—fine blue.

o—11^h 12^m S 3° 37'—yellow. Mira, the celebrated variable, from 2 to 7 mag., sometimes to invisibility, in 331^d 15^h 7^m according to H. Its maximum brightness and period are not always the same: Argelander has shewn the probability of regular alternations in the latter. One of the most interesting problems of modern astronomy is the question whether the irregularities of variable stars may not be, like the planetary perturbations, phases of some general law.

Double Stars.

γ—11^h 36^m N 2° 39'—3, 7—2·6"—pale yellow, lucid blue, 1831-1843. Dembowski 7 olive-green, 1854 [tawny, 1850]. Σ gives them common proper motion.

37—1^h 7^m S 8° 41'—6, 7·5—51"—white, light blue or dusky. Σ common proper motion. $n\ p$ lies another pair—8, 10—20"—yellow, violet. 2° $p\ \theta$.

146 P O—0^h 34^m S 5° 7'—6·5, 9—58"—pale topaz, violet.

66—11^h 6^m S 2° 9'—7, 8·5—15"—pale yellow, sapphire blue. Common proper motion? 1½° p Mira, a little n.

113 P O—0^h 27^m S 5° 19'—7, 9—20"—cream yellow, smalt blue. [9 over-rated?]

61—1^h 57^m S 1° 1'—7, 11—39"—pearly white, 1834

[CETUS]

[pale orange, 1850], greenish. I found the attendant very obvious to the averted eye. Followed by 218 Σ—7, 8·5—4·6″—white, blue. 3° *s* a little *f a* ⅓.

[χ—1ʰ 42ᵐ   S 11° 30′ forms a fine pair with 182 P I—5, 7·5—pale yellow, bluish. Closely *s p ζ*.]

[About 1ʰ. 35ᵐ   S 8°, is a beautiful 8 mag. pair, *s* a little *p* 167 P I, 6 mag.]

Nebula.

77 M—11ʰ 36ᵐ   S 0° 36′.   1° *f* δ, a little *s*. Small, faintish; very near a 9 mag. star.   ♅ thought it at least 910 times more distant than 1 mag. stars !

CLYPEUS (or SCUTUM) SOBIESKII.

This asterism, which worthily associates the memory of the Polish hero with the most brilliant part of the Galaxy visible in our latitudes, is full of splendid telescopic fields: and the very ground of the Milky Way seems here resolvable.

Clusters and Nebula.

24 M—XVIIIʰ 10ᵐ   S 18° 27′. Magnificent region, visible to the unaided eye as a kind of protuberance of the Galaxy. It is accompanied by two little pairs. 2° *n* of μ ♐.

16 M—XVIIIʰ 11ᵐ   S 13° 50′. Grand cluster.

18 M—XVIIIʰ 12ᵐ   S 17° 11′. Glorious field in a very rich vicinity. *s* lies a region of surpassing splendour.

26 M—XVIIIʰ 38ᵐ   S 9° 32′. Coarse cluster.

17 M—XVIIIʰ 13ᵐ   S 16° 15′. The " horse-shoe ", nebula, visible in finder, 1° *n* of 18 M, described by Smyth as " a magnificent, arched, and irresolvable nebulosity,—in a

splendid group of stars." Well has he observed, " the wonderful quantity of suns profusely scattered about here would be confounding, but for their increasing our reverence of the Omnipotent Creator, by revealing to us the immensity of the creation."

COMA BERENICES.

A gathering of small stars, which obviously at a sufficient distance would become a nebula to the naked eye.

Double Stars.

12—XII^h 15^m N 26° 37'—5, 8—66"—straw-yellow, rose-red. 1½° s p 16, the "lucida" of the constellation. [This latter is beautifully placed in a little triangle of 8 or 9 mag. stars. It is marked 3 mag. in Map of S. D. U. K., but was not larger than 5 or perhaps 4 mag. 1852.]

24—XII^h 28^m N 19° 9'—5·5, 7—21"—orange, emerald.

2—XI^h 57^m N 22° 14'—6, 7·5—3·6"—pearly white, lilac.

202 P XII—XII^h 45^m N 19° 56'—7·5, 8—16"—both white. 2° s of 35.

[17—about 1° s a little f 16, forms a fine pair with 96 P XII; colours somewhat different. The smaller star, by averted vision, seems more surrounded than the other with scattered light.]

Nebula.

[85 M—about XII^h 18^m N 19°. Fair specimen of the many nebulæ in this region: midway from 24 towards 11, the nearest conspicuous star p, a little s.]

CORONA BOREALIS.

A constellation resembling more than usual the object whose name it bears.

Double Stars.

ζ—xv^h 34^m N 37° 6′—5, 6—6·1″—bluish white, smalt blue.

ν²—xvi^h 17^m N 34° 2′—5, 12 [10, 1850, 1855]—137″— pale yellow, 1838 [deep yellow, 1855], garnet, forms a fine group with ν¹—5—deep yellow, and a 6 mag. grey star f.

η—xv^h 17^m N 30° 48′—6, 6·5—1″ more or less, less 1859—white, golden yellow. One of ♅'s severest tests. Smyth divided it, 1832, but could not always even elongate it, 1842. It may be interesting to look at so wonderful an object as a pair of suns revolving in the brief period of 44 years (or, according to Villarceau, 67 years,) even though we can only see them closed up into a single star. Visible to naked eye, a little out of the curve of the coronet.

σ—xvi^h 9^m N 34° 13′—6, 6·5—1·8″, widening, 1843. (Fletcher 2·3″, 1851)—creamy white, smalt blue, 1830—1843. South 6·5 " certainly not blue; it differs very little from the large star in colour," 1825. Σ 6·5 alba, 1836. Dembowski yellow, yellow sometimes ashy, 1854–5; white, ashy, 1856; red, doubtful blue, 1857. [6·5 sometimes ruddy, sometimes bluish, 1850, 1855, with more than ½ mag. difference: Dembowski rated them 5·3, 6·5, 1854–5.] Binary: probable period not less than 560 years; followed at 44″ by a little blue star, 11 mag. Smyth: 15 or 20 mag. South. 1825, not visible with

more than 92 of a very fine 7 feet achromatic, 5 inches aperture [readily seen with 80, 144, 250 of 5½ feet achromatic].
2° *p v.*

CORVUS.

This small constellation contains several conspicuous stars.

Double Stars.

β—xII^h 27^m S 22° 37'—ruddy yellow, 1831 [pale yellow,
1852]. The companions in the Bedford Catalogue are distant:
but it is inserted to be watched for variation.   Smyth found,
1831, that though possessing no Arabic name, and lettered β
by Bayer, it was unquestionably the brightest in the constellation.   ♄, 1783, gave the order of brightness γ δ β a ; 1796,
γ β δ a, with but little difference between them : I found the
order, 1852, 1854, 1859, γ δ β a.

δ—xII^h 23^m S 15° 44'—3, 8·5—24''—pale yellow, purple.

CRATER.

Like Corvus, an appendage to Hydra.

Double Stars.

17—xI^h 25^m S 28° 30'—5·5, 7, 1833—10''—lucid white,
violet. Relatively fixed, common proper motion. [Only ½
mag. difference, 1852.]

39 P XI—xI^h 13^m S 6° 8'—8·5, 9—8''—both bluish white.
[A 10 mag. pair in field.   About 45' *p* is a fine pair,—7, 9—
white, bluish or lilac.]

CYGNUS.

This fine cruciform constellation occupies a prominent position in the Galaxy, and nothing can be more magnificent than the low-power fields which it contains. Its principal star, Deneb, has no perceptible parallax, or proper motion; so far deserving the title, usually very inappropriate, of a "fixed star"; hence we must infer amazing distance, and magnitude surpassing possibly that of Arcturus, Wega, or even Sirius itself.

Double Stars.

β—XIX^{h} 25^{m} N 27° 40'—3, 7—34"—topaz yellow, sapphire blue. One of the finest in the heavens. I have seen the colours beautifully by putting the stars out of focus. Smyth observes that they are actually different, not, as may sometimes be the case, complementary, from mere contrast; an effect which is seen when the bright yellow light of a lamp makes the Moon appear blue, and which Schmidt witnessed to a remarkable degree at his observatory on Vesuvius during the great eruption of 1855, when the sky was as green as bottle-glass, and the Full Moon a lively green, through openings in red clouds of smoke and steam. A similar result may take place with some double stars, but not with all, as is proved by hiding the larger star behind a bar in the field. On this account the presence of artificial light is objectionable in observations of the colours of stars.

o^{2}. Triple—XX^{h} 9^{m} N 46° 19'—4, 7·5, 5·5(o^{1})—107"—338"—orange, blue, 1838 [white, 1850], blue. Smyth found both the smaller stars cerulean blue when the larger one was concealed. Glorious field.

μ. Triple—XXI^h 38^m N 28° 7′—5, 6, 7·5—5·4″, 217″—white, two blue. Σ finds common proper motion in the close pair.

χ—XIX^h 41^m N 33° 25′—5, 9—26″—golden yellow, pale blue. Relatively fixed, common proper motion. Closely f, a little s, is 295 P XIX, variable, often incorrectly called χ : period about 7 months; sometimes reaching 5 mag.; very red, like many variables, according to Schmidt. I have not looked for it.

ω^3—XX^h 27^m N 48° 45′—5, 10—55″—pale red, grey. 199 P XX—7, 9·5—61″—white, pale blue, precedes it, making a fine group. Two very faint stars, which direct vision with my 3$\frac{7}{10}$ inches aperture would just reach, form a trapezium with ω^3.

61 —XXI^h 0^m N 38° 3′—5·5, 6—16″—yellow, deeper yellow. One of the most interesting objects in the sky : these suns not only form a binary system revolving in upwards of 540 years; they not only travel through space at a rate which would, as Humboldt observes, carry them across 1° in 700 years ; but they were the first of the host of heaven to reveal to Bessel the secret of their distance.* This is probably 657,700 times that of the Earth from the Sun—itself 95,000,000 miles—a space so vast, that light, which reaches us from the Sun in 8^m, employs more than 10 years to traverse it : we see these stars,

---

* This grand result was obtained with the Königsberg Heliometer. The instrument absurdly so termed, as if it were peculiarly intended to measure the *Sun*, is an achromatic of which the object-glass is cut into two halves; a slight displacement of these produces a double image, and affords the means of very accurate measurement. There is a fine heliometer at Oxford, 7$\frac{1}{3}$ inches aperture.

therefore, not as they are now—for of their present existence we have no information—but as they were 10 years ago.*

How vast must be the dimensions of this great Universe! What a temple for the Creator's glory! "All the whole heavens are the LORD's"—those heavens are crowded with millions upon millions of stars; and of all that countless multitude, millions, probably, for one, are at a distance incalculably exceeding that of 61 Cygni!

ψ—XIX^h 52^m N 52° 4'—5·5, 8—3·5"—bright white, lilac.

52—XX^h 40^m N 30° 13'— 5·5, 9·5—7"— orange, blue. 3° s of ε.

429 P XX—XX^h 54^m N 49° 55'—6, 7·5 — 2·1" — silvery white, pale grey.

278 P XIX—XIX^h 41^m N 34° 40'—6, 8—39"—straw-colour, smalt blue. 1$\frac{1}{4}$° n of χ.

49—XX^h 35^m N 31° 49'—6, 9—3·2"—golden yellow, blue. 2° s p ε.

16—XIX^h 38^m N 50° 12'— 6·5, 7, 1834 [not much inequality, 1850-1]— 37"—both pale fawn colour. Σ thinks they have a common motion through space. Within 1° $n f \theta$.

1 P XXI—XXI^h 3^m N 29° 38'—6·5, 9—3·5"—dull white, pale lilac. 1$\frac{1}{2}$° p ζ.

452 P XX—XX^h 57' N 38° 58'— 7, 11—17"—deep yel-

---

* From the successive transmission of light results the extraordinary fact that the aspect of the whole heavens is of unequal date, each star having its own time of "light-passage" to our eyes, and those times immensely differing, so that there is no impossibility in Humboldt's magnificent assertion, "much has long ceased to exist before the knowledge of its presence reaches us; much has been otherwise arranged." H thinks the little components of the Galaxy may be distant 2,000 years!

low, emerald [11 obvious, 1850, several hours past meridian.]

276 P XIX — XIX^h 41^m N 35° 45′— 8, 8·5 — 15″—both white.

149 P XIX — XIX^h 23^m N 36° 15′—8·5, 9 — 7″—white, pale blue.   Fine field: closely f 4, a 5 mag. star.

(6 — XIX^h 7·5^m N 49° 31′— 6, 6 — 10·5″— both yellow.)

(2705 Σ—XX^h 30·6^m N 32° 44′—6·5, 8—3″—yellow, blue.)

(2760 Σ—XX^h 59·5^m N 33° 25′—7,8—12·7″—white, ashy.)

(2747 Σ—XX^h 55·5^m N 36° 58′—7·7, 8—45″—both white.)

[These 4 pairs from Σ must be very fine. I have not seen them.]

(Variable — In the field with θ, XIX^h 32^m N 49° 50′, Pogson has discovered a variable star, with maximum nearly 7 mag. and period about 300^d.   This I have not looked for.)

Clusters.

39 M—XXI^h 27^m N 47° 49′.   Brilliant group and vicinity.

[Near 36, 29, and 28, three 5 mag. stars, about 3° s of γ, a little p, is a superb region.]

[32 — about XX^h 11^m N 47° 10′— 5 — dull orange, is in a fine field.]

DELPHINUS.

The leaders of this little, compact, fish-like constellation, α and β, are distinguished by names which, even among the multifarious disfigurements of Oriental words so abundant in the heavens, are pre-eminently strange, *Svalocin* and *Rotanev*. The former Smyth has justly characterized as " cacophonous and barbaric," and says that " no poring into the black letter versions of the Almagest, El Battáni, Ibn Yúnis, and other authorities, enables one to form any rational conjecture as to the mis-reading, mis-writing, or mis-application, in which so strange a metamorphosis could have originated." And of *Rotanev* he observes, "the which putteth derivation and etymology at defiance." Where so eminent and accomplished a scholar and antiquarian has not succeeded, it would seem presumptuous to offer a solution, but that accident is sometimes more fortunate than study; and if the following is not after all the right key, it certainly is a marvel that it should open the lock so readily. The letters of these strange words, reversed, form NICOLAUS VENATOR, in which it is easy to recognize a Latin version of the name of NICCOLO CACCIATORE, assistant at the Palermo Observatory, in the Catalogue emanating from which these stars are so denominated.*

Double Stars.

γ—xx^h 40^m N 15° 37'—4, 7—12"—yellow, light emerald, 1831—1839. ♅ both white, 1779; hence Σ suspects change. H and South, white, yellowish, 1824.

178 P XX—xx^h 25^m N 10° 48'—7·5, 8—14"—both white. Closely $s\,p\,\epsilon$.

[a and 247 P XX form a fine combination; pale yellow, pale lilac.]

[β and ζ, two fine yellow stars, have a beautiful little 8 mag. triangle between them.]

[θ is in a beautiful field.]

Nebula.

103 ♅ I—xx^h 27^m N 6° 57'. Resolved into stars by ♅.

* Cacciatore died in 1841, from the effects of cholera.

DRACO.

A long, winding constellation, always above the horizon ; in consequence of which its stars, like all others in the vicinity of the pole, appear at different times *entirely reversed* in relative position. A careful attention to p and f, that is, to the direction of apparent motion through the field, is in these cases required to prevent mistakes in identification. Here are many fine pairs.

Double Stars.

μ — XVII^h 2^m N 54° 40′—4, 4·5—3·3″—both white. Binary, revolving in perhaps 600 years.

ν^1 — XVII^h 29^m N 55° 17′— 5, 5 — 62″— both pale grey. Probably common proper motion.

39. Triple—XVIII^h 22^m N 58° 43′—5, 8·5, 7—3·3″, 89″ —pale white, light blue, ruddy.

o—XVIII^h 49^m N 59° 13′—5, 9—30″—orange yellow, lilac.

ψ^1—XVII^h 44^m N 72° 13′—5·5, 6—31″—both pearly white, 1838. Σ albæ, 1832 (who gives them proper common motion). Dembowski whitish yellow, ashy yellow, 1856. [yellow, lilac, in evident contrast, 1850.]

40 and 41 — XVIII^h 11^m N 79° 59′— 5·5, 6 — 20″— both white, 1839 [yellow, paler yellow, 1856 : grouped finely with a smaller lilac star].

ε — XIX^h 49^m N 69° 55′—5·5, 9·5 — 3·1″— light yellow, blue : 9·5 variable ? H and South found it very difficult ; it was easy to me, and apparently large for the assigned magnitude. The contrast is very pleasing.

17. Triple — XVI^h 33^m N 53° 12′—6, 6·5, 6—3·8″, 90″—

[DRACO]

pale yellow, faint lilac, white. Σ (1833) considered the close pair both white, and alternately variable.

147 P XVII — xvɪɪʰ 26ᵐ   N 50° 59′—8, 8·5—3·2″—pale white, ruddy, 1836. Dembowski, both white, 1854.   1½° s of β, the Dragon's eye.

[46—xvɪɪɪʰ 39ᵐ   N 55° 20′—5, 9—full yellow, clear blue; a fine contrast.]

[About 4° from α Drac. towards α Urs. Maj. is a striking pair—6, 6·5—yellowish white, white.]

[9 ? Triple — xɪɪʰ 54ᵐ   N 67° 30′— 6·5, 6·5, 8 — two yellow, bluish.]

(2218 Σ—xvɪɪʰ 39·2ᵐ   N 63° 40′—6·5, 8·5—2·5″—yellowish, blue, must be a beautiful pair. I have not looked for it.)

Nebula.

37 ♅ IV—xvɪɪʰ 59ᵐ N 66° 38′. Planetary : a very curious object. I found it much like a considerable star out of focus; very bright for its class. ♅ gave it 35″ diam. I could see but 15″ or 20″, and could not well make out the pale blue colour ascribed to it by Smyth. Nearly half way between Polaris and γ Drac.; in the pole of the Ecliptic.

EQUULEUS.

This little asterism is easily recognized by the clustering of its stars, and its position relative to Pegasus. It contains a few good objects, and a low power will exhibit many interesting fields.

Double Stars.

ε—xxʰ 52ᵐ N 3° 46′—5·5, 7·5—11″—white, lilac. 5·5 is very closely double and binary. This is 1 on the Map.

λ—xx^h 55^m N 6° 38' — 6, 6·5 — 2·6" — both white.   A beautiful close pair.

376 P XX—xx^h 49^m N 4°0'—6, 8—1·8"—orange, purple, 1833.   Dembowski white, blue, 1854.   [Elongated, 80 : clearly divided, 144.]   Second star p ε (or 1).

355 and 356 P XX—xx^h 46^m N 6° 48'—8·5, 8·5—40" —both white.

[γ and 6,—5, 6—pale yellow, white, are a striking pair.]

[Between α and δ is a wide 7 mag. pair, both white.]

[δ is followed by 3 little stars, singularly arranged in a straight line.]

ERIDANUS.

An asterism winding down to S. horizon, its *lucida* being out of sight in our latitudes.

Double Stars.

32—III^h 47^m S 3° 22'—5, 7—6·8"—topaz yellow, seagreen, or flushed blue.

39—IV^h 8^m S 10° 36'—5, 11 [perfectly easy]—7·1"— full yellow, deep blue.

62—IV^h 50^m S 5° 24'—6, 8—64"—white, lilac. 2$\frac{1}{2}$°p β.

98 P III—III^h 30^m N 0° 8'—6·5, 9—5·9"—yellow, pale blue.

55—IV^h 37^m S 9° 4'—7·5, 7·5—10"—both yellowish white.

Nebula.

26 ♅ IV—IV^h 8^m S 13° 6'.   Planetary: bright and round with low powers, but not bearing magnifying well with

my aperture.　Lassell describes it as the most interesting and extraordinary object of the kind he had ever seen; an 11 mag. star standing in the centre of a circular nebula, itself placed centrally upon a larger and fainter circle of hazy light.

GEMINI.

The leading stars, in the heads of the two figures, are well known, but it requires a little attention to the globe or map to make out the whole constellation.

Double Stars.

α　(Castor)—VII^h 26^m　N 32° 12′—3, 3·5—4·9″—bright white, pale white.　H calls it "the largest and finest of all the double stars in our hemisphere;" its rapid motion first fully convinced ♁ of the existence of binary systems.　Period not quite certain: H gives it 253 years; Smyth, 240 years. Excellent test for a small telescope.

ε—VI^h 35^m N 25° 16′—3, 9·5—111″—brilliant white, cerulean blue, 1831 [3 strongly yellow, 1849].

μ—VI^h 14^m N 22° 35″—3, 11—80″—crocus yellow, bluish. The tint of 3 appeared to me very fine.

γ—VI^h 30^m　N 16° 31′—brilliant white.　[A low power shews a pretty field of minute stars, appearing to radiate from it in every direction.]

38—VI^h 47^m N 13° 21′—5·5, 8—·5·8″—light yellow, purple : perhaps binary.　Σ finds mags. vary.

15—VI^h 19^m　N 20° 52′—6, 8—33″—flushed white, bluish.

20—VI^h 24^m　N 17°53′—8, 8·5—20″—topaz yellow, cerulean blue.　Field fine : 1½° $n\,p\,$ γ.

61—VII^h 19′ N 20° 32′—7·5, 9—60″—deep yellow (1835), yellowish. [7·5 white, and no companion larger than 11 mag., 1852, 1855.] This pair points to another $n\,p$, 48 ♓ III—8, 9—6·5″—blue, bluish. 2$^{\cup}$ $s\,f\,\delta$.

[A fine pair—7·5, 9—red, blue, is about 40′ n of 63, a 6 mag. star with a minute attendant, which is 2° f a little s from δ.]

Cluster and Nebula.

35 M—VI^h 0^m N 24° 21′. Beautiful and extensive region of small stars, a nebula to naked eye: how differently Lassell's 24 inch mirror shews it, his own words will tell: — " A marvellously striking object. No one can see it for the first time without an exclamation . . . the field of view, 19′ in diameter and angular subtense 53½°, is perfectly full of brilliant stars, unusually equal in magnitude and distribution over the whole area. Nothing but a sight of the object itself can convey an adequate idea of its exquisite beauty." Smyth observes* that the stars form curves, often commencing with a larger one: (see note, p. 231.) It lies between ε Ⅱ and ζ ♉, a little n.

45 ♓ IV—VII^h 21^m N 21° 12′. ♓ observed this object as a 9 mag. star, "with a pretty bright nebulosity, equally dispersed all around: a very remarkable phænomenon." H describes it as an 8 mag. star, "exactly in the centre of an exactly round bright hemisphere 25″ in diameter." Smyth, who rates it 7·5 mag., says he " could only bring it to bear as a burred star." I was so much surprized at the result in my inferior telescope, that I cannot help supposing some temporary impediment to distinct vision at Bedford, for on coming accidentally across it in 1850, I found such a conspicuous nebulosity that I thought it was either damp on the eye-

lens, or a telescopic comet; and in 1852 I entered it as a
" bluish nebulosity quite like a telescopic comet." The Earl
of Rosse sees a marvellous object: a star surrounded by a
small circular nebula, in which, beside the star, is a little dark
spot; this nebula is encompassed, first by a dark, then by a
luminous ring, very bright, and always flickering. It lies 2°
$s f \delta$.

HERCULES.

Some very fine telescopic objects mark this constellation.

Double Stars.

a—XVII^h 8^m N 14° 33'—3·5, 5·5—4·5"—orange, emerald
or bluish green. Smyth calls it a " lovely object, one of the
finest in the heavens." ♅ makes 3·5 variable, 3 to 4 mag.;
Σ finds 5·5 variable, 5 to 7 mag. Apparently stationary.

ρ—XVII 19^m N 37° 17'—4, 5·5—3·7"—bluish white, 1839,
pale emerald. Dembowski 4 reddish white, 1853, 1855.

δ—XVII^h 9^m N 25° 0'—4, 8·5—25"—greenish white, grape
red, 1830–7–9. Σ viridis, albacinerea. Fletcher yellow, red,
1851. Dembowski yellow, blue, 1854–5 ; white, blue, 1855–
6. [pale yellow, bluish green, 1850.] Probably binary.

μ—XVII^h 41^m N 27° 48'—4, 10—30"—pale straw colour,
cerulean blue. Apparently relatively fixed, with common
proper motion : if so, their real sizes must widely differ.
10 is now found very closely double.

λ—XVII^h 25^m N 26° 13'—4·5—inserted for its curious
colour, to my eye deep, dull orange. Towards this point the
whole solar system, according to ♅ and Argelander, is moving.

95—XVII^h 56^m N 21° 36'—5·5, 6—6·1"—light apple-green,
cherry-red. Smyth observes that " this beautiful object pre-

sents a curious instance of difference in colour between components so nearly equal in brightness." It is indeed an
extremely pretty pair, and well worth the search; mark, on
the Map, its configuration with α Herc. and α Oph.

κ^1 and κ^2—xvi^h 2^m N 17° 25'—5·5, 7—31"—light yellow, pale garnet.

23—xvi^h 18^m N 32° 40'—6, 9—36"—white, violet.   1½°
s of ν Cor.

200 P XVII—xvii^h 35^m N 24° 35'—6·5, 9—16"—topaz
yellow, purple.   [A good low-power field follows.]

100—xviii^h 2^m N 26° 5'—7, 7—14"—both pale white.

300 P XVII—xvii^h 50^m N 18° 21'—7·5, 8—2.5"—both
lucid white.

46—xvi^h 40^m N 28° 37'—7·5, 10—5·1"—pale white, sky
blue.

[A conspicuous pair—6, 7—flushed white, bluish, lies about
2½° n f from β towards 51.]

[125 and 126 P XVI—xvi^h 28^m N 17° 20'—7, 8.]

[A pretty 8 mag. pair, both white, is 25' n f 5, 6 mag.,
about xv^h 54^m N 18° 20'.]

Nebulæ.

13 M—xvi^h 37^m N 36° 43'.   Glorious resolvable nebula,
lying ⅓ from η towards ζ; the finest of its class; just visible to
naked eye.   Halley discovered it in 1714; M was *sure* it contained *no stars;* but it is spangled with glittering points in a
5½ feet achromatic, and becomes a superb object in large
telescopes.   Smyth calls it "an extensive and magnificent
mass of stars, with the most compressed part densely compacted
and wedged together under unknown laws of aggregation:"

in Secchi's achromatic the outliers, inconspicuous in ordinary instruments, fill a field of 8'. The aspect of this collection of innumerable suns is enough to make the mind shrink with a sense of the insignificance of our little world. Yet the Christian will not forget that, as it has been nobly said, HE took of the dust of this earth, and with it HE rules the universe!

The neighbourhood is beautiful with a low power.

92 M — XVII^h 13^m N 43° 17'. Very fine nebula, not equal to 13 M, but intensely bright in centre. In ♅'s reflectors, a brilliant cluster of 7' or 8' diameter.

5 N, Σ—XVI^h 39^m N 24° 3'. Planetary nebula, 8'' diameter, discovered by Σ; I found it very bright, but small and not sharply defined; exactly like a star out of focus, bearing my high powers well. Secchi resolves it with a power of 1,500. Rather more than 1° sp 51, 5 mag.

50 ♅ IV—XVI^h 43^m N 46° 8'. Planetary nebula, faint in my achromatic, but beautifully grouped in a triangle with two 6 mag. stars.

HYDRA.

A very lengthy and not very interesting constellation to the unaided eye; but containing a good deal of telescopic work.

Had I ever been able to find H's 8 mag. star, "scarlet, almost blood-colour, a most intense and curious colour," which follows α 42·5^s, 1' s, I should have included it in the list; but I have failed repeatedly, probably from want of light. I have mentioned it, however, for the sake of the possessors of larger instruments. H and Hind mention several similar instances of scarlet or ruby stars.

Double Stars.

ε —VIII^h 39^m N 6° 56'—4, 8·5—3·6"—pale yellow, purple. Dawes suspected it of being binary. Smyth suggests a period of about 450 years.

10 — XIV^h 38^m S 24° 51'— 5·5, 7·5 — 9·8"— pale orange, violet. Probably binary.   54 in Map, after ♅. Very near horizon.

τ^1— IX^h 22^m S 2° 9'—5·5, 8·5—65"— flushed white, lilac.

108 P VIII —VIII^h 28^m N 7° 6'—6, 7—11"—pale yellow, rose tint. Finely grouped with other stars, 1° $n\,p\,\delta$ (in the head).

160 P VIII—VIII^h 38^m S 2° 6'—7, 8—4·9"—silvery white, smalt blue.

17—VIII^h 49^m S 7° 26'—7·5, 7·5—4·5"—both white.

65 and 64 P IX — IX^h 16^m N 4° 6'— 8, 9 — 22"— both white, 1836 [reddish white, grey or bluish, 1851].

159 P X—X^h 41^m S 14° 53'—8,9—32"—pale white, light blue. I found this a guide to a much finer object, a star followed by a pair, 1474 Σ — X^h 39·1^m S 14° 21'—all 7 to 8 mag., and nearly in a line; 1½° n of ν Crat.

[A pretty open 8 mag. pair follows γ, about 1° distant.]

(Variable. 94 P XIII — XIII^h 21^m S 22° 30'. Argelander gives it an irregular period. Pogson says it may attain 4 mag. I have not seen it.)

Nebula.

27 ♅ IV — X^h 18^m S 17° 57'. Planetary nebula, 2° s from μ, resembling Jupiter, as Smyth says, in size, equable light, and colour. I found it bright, a little elliptical $n\,p$, $s\,f$, of a steady pale blue light, bearing high powers. ♅ did not re-

solve it. Secchi, whose beautifully defining glass accomplishes marvels with 1,000, finds it an unique object; within a circular nebulosity it contains two clusters connected by two semicircular arches of stars forming a sparkling ring, with one star on the hazy ground of the centre.

LACERTA.

A small and not distinctly marked asterism.

Double Stars.

8². Quadruple — xxii^h 30^m N 38° 55'—6·5, 6·5, 11, 10 —two nearest, 23"—two white, greenish, blue.

65 P XXII — xxii^h 13^m N 37° 4'— 6·5, 9 — 15"—pale white, livid. Closely *f* 1, 5 mag.

Clusters.

75 ℍ VIII—xxii^h 10^m N 49° 11'. Fine cluster, quickly followed by a beautiful field with 3 pairs.

[7—xxii^h 24^m N 49° 25', points out a noble field. 4, a 5 mag. star, 1ᵛ *s p* 7, is a fine object, deep orange, with a blue attendant, in a rich field.]

LEO.

A fine constellation, the fore part of which is marked to the naked eye by a *sickle* composed of conspicuous stars. At the bottom of the handle, and very nearly in the pathway of the Sun, is the leader, Regulus, the Lion's Heart, the first in the following list.

Double Stars.

α — x^h 1^m N $12°$ $39'$ — flushed white.   Σ and Secchi think a distant 8·5 attendant, which affords a beautiful contrast, is moving with it through space.

β — xi^h 42^m N $15°$ $21'$ — $2·5$, 8 — bluish, dull red.   Very wide.

γ — x^h 12^m N $20°$ $33'$ — 2, 4 — $2·8''$ — bright orange, greenish yellow; ♅, white, reddish white.   Binary? perhaps period of 1,000 years.   A very fine object.

ι — xi^h 17^m N $11°$ $18'$ — 4, $7·5$ — $2·5''$ — pale yellow, light blue.   Binary.   I just divided it with 80.

54 — x^h 48^m N $25°$ $30'$ — $4·5$, 7 — $6·2''$ — white, grey.

90. Triple — xi^h 27^m N $17°$ $34'$ — 6, $7·5$, $9·5$ — $3·5''$, $59''$ — silvery white, purplish, pale red.

49 — x^h 28^m N $9°$ $22'$ — 6, 9 — $2·5''$ — silvery white, pale blue.   [Well seen with 80.   Good test; $1°$ $s f \rho$.]

6 — ix^h 24^m N $10°$ $20'$ — 6, $9·5$ — $38''$ — flushed yellow, pale purple, 1851 [deep orange, green, 1851].

88 — xi^h 25^m N $15°$ $9'$ — 7, 9 — $15''$ — topaz yellow, pale lilac.   Σ gives them common proper motion.

83 — xi^h 20^m N $3°$ $47'$ — 8, 9 — $30''$ — silvery white, pale rose.   Relatively fixed, common proper motion.   Closely $n p$ τ, 4 mag., itself forming with 71 P XI, 8 mag., a fine pair — yellow, violet.

67 P X — x^h 18^m N $9°$ $29'$ — 8, $9·5$ — $3·5''$ — white, pale blue.   Closely f 44 — 6 — orange, which is $2°$ $p \rho$.

239 P X — x^h 59^m N $7°$ $53'$ — 8, $11·5$ — $8·2''$ — topaz yellow, cerulean blue. $11·5$ very difficult, 80; steady, 144, so as to be a good comparative test, and easily found; closely f χ, but not quite so near it as a larger star further n in the field.

(Variable. 176 P IX (alias 420 Mayer)—IX^h 40^m N 12°
5′. I have not seen this star, which lies a little f 19, 7 mag.,
and is the second of two in the field. Pogson gives it 5 mag.
at maximum ; diminishing to 10 mag.; with a very irregular
period of about 312^d. Hind says, " it is one of the most fiery
looking variables on our list—fiery in every stage from maximum to minimum, and is really a fine telescopic object in a
dark sky, about the time of greatest brilliancy, when its colour
forms a striking contrast with the steady white light of the
6 mag. a little to the n." It is R ♌ in Argelander's nomenclature for variables. Many of these mysterious objects are
too minute for general observation : those who can follow them
find them endless sources of wonder. Many, though not all,
are intensely red; the times of increase and decrease are often
strangely unequal; some are hazy, others quite sharp, at their
minimum. Baxendell found one (U II) hazy at its maximum : this star increased, Nov. 1858, at the rate of 1½ mag.
per day ! Observation is adding constantly to their number,
but throws no light, at present, upon their constitution.)

Nebulæ.

66 and 65 M—XI^h 13^m N 13° 46′. Two rather faint objects, elongated, but in different directions, in a low-power
field, with several stars. 66, s, rather the larger and brighter.
Between ι and θ, a little f.

18 ♅ I—X^h 41^m N 13° 22′. Two faint nebulæ. p much
the larger and brighter, with stellar nucleus. Smyth mentions
a neat little pair $n f$ [well seen with 80]. These are among
the nebulæ in a round patch of 2° or 3°, in a region containing
few stars.

LEO MINOR.

This small constellation contains several of ♅'s nebulæ, but too faint for ordinary instruments. The following may be specified.

Nebula.

200 ♅ I — viii^h 44^m N 33° 56'. ♅ calls this a very beautiful object, 8' long, 3' broad : Smyth saw a splendid centre. I found it scarcely worth the search ; but it lies in a very fine neighbourhood, a little p the most n group of ♋.

LEPUS.

A little asterism under the legs of Orion ; so near the horizon that it can only be well seen on the meridian, and opportunities must not be thrown away.

Double Stars.

γ — v^h 39^m S 22^u 30'— 4, 6·5 — 93"— light yellow, pale green, 1832 [pale yellow, garnet, 1851].

ι — v^h 6^m S 12° 2'—4·5, 12 — 15"—white, pale violet. I found 12 certain with 80, a glimpse star with 144 ; this seems an exemplification of Smyth's remark, that among very minute stars, the smallest sometimes shine with a keener light than those of larger apparent magnitude.

κ — v^h 7^m S 13° 6'—5, 9—3·7"—pale white, clear grey.

(Variable — iv^h 53^m S 15° 2'. Discovered by Hind, 1845, who describes it as "of the most intense crimson, resembling a blood-drop on the black ground of the sky ; as regards depth

⌊LEPUS⌋
of colour, no other star visible in these latitudes could be compared to it." It was between 6 and 7 mag. up to 1854. In
1855 Schmidt considered it rapidly gaining light, but losing
colour. I have not seen it.)

Nebula.

79 M — v^h 19^m S 24°39′. Tolerably bright with my 64,
blazing in centre; higher powers shew it mottled : a beautiful
cluster of stars in ♓'s 20 feet reflector, nearly 3′ diameter.
4° s a little p β, closely f a 6 mag. star.

LIBRA.

S. declination combines with long days and late sunsets to
give trouble in looking for the objects in this constellation,
which are, however, well worth the pains.

Double Stars.

β—xv^h 9^m S 8° 52′—2·5. Inserted for its beautiful pale
green hue, very unusual among conspicuous stars : dark green,
like dark blue, is unknown to the naked eye.

α^2 and α^1—xiv^h 43^m S 15° 27′—3, 6—229″—pale yellow,
light grey.

51—xv^h 57^m S 10° 59′—4·5, 7·5—7·2″—bright white,
grey. Really triple, a 5 mag. star lying close to 4·5 : all three
seem moving. An 8 mag. pair, s a little f in the field (n p in
the Bedford Catalogue) forms a beautiful group with it. 51 is
ξ ♎ of Flamsteed and Σ.

212 P XIV—xiv^h 49^m S 20 °46′—6, 8 — 10″—straw-
coloured, orpiment yellow. Triple in first-rate telescopes.

91 P XV—xv^h 24^m S 19° 41'—7·5, 9 [very small?]—12" —bluish white, smalt blue.

62 P XIV—xiv^h 15^m S 7° 7'—8, 8—5·2"—both silvery white. 2½° $s f \iota$ ♍.

14 P XV—xv^h 7^m S 17° 54'—8, 9—48"—silvery white, pale grey.

Cluster.

5 M—xv^h 11^m N 2° 37'. Beautiful assemblage of minute stars, greatly compressed in centre. M assured himself that it contained no stars. ♅ with 40 feet reflector counted about 200. It lies closely np 5 Serp., 5 mag.

LYNX.

A troublesome constellation, excepting with an equatorial mounting, as there are few conspicuous leaders among a number of tolerably considerable stars, which do not occur in the Map of the Society for the Diffusion of Useful Knowledge, and are puzzling in the finder. The beauty of the pairs which it contains will, however, reward the observer's perseverance.

Double Stars.

38—ix^h 10^m N 37° 24'—4, 7·5—2·8"—silvery white, lilac.

12. Triple—vi^h 34^m N 59° 35'—6, 6·5, 7·5—1·6", 8·9"— white, ruddy, bluish, 1832, 1839 [ruddy, 1854]. I elongated this with 80, divided it at times with 144. Binary, with period of perhaps 700 years.

301 P VI—vi^h 55^m N 52° 58'—6, 6·5—3"—both white. [little difference in mag.—so Σ.]

41—ix^h 19^m N 46° 13'—6·5, 8·5—87"—both bluish,

P

[LYNX]

1832 [deep yellow, lilac, 1852]. Smyth thinks P's mag. 6·5 under-rated. A 10 mag. star forms a triangle.

39—IX^h 13^m N 50° 8'—6·5, 9—6·2"—lucid white, sapphire blue. 2½° $s\ p\ \theta$ Urs. Maj.

19. Triple—VII^h 11^m N 55° 33'—7, 8, 8—15", 215"— white, two plum coloured.

20—VII^h 12^m N 50° 24'—7·5, 7·5—15"—both silvery white.

174 P VI—VI^h 32^m N 59° 35'—7·5, 10—4"—bright white, blue. Σ thinks 10 variable, and this, Smyth observes, " awakens considerations of peculiar interest, it having been surmised that certain small acolyte stars shine by reflected light." Exactly p 12.

131 P VIII—VIII^h 34^m N 49° 22'—8·5, 8·5—9·8"—both white. About 2^v $n\ p\ \iota$.

[40—IX^h 11^m N 35° is a fine 4 mag. star, orange, with an 8 or 9 mag. companion, violet.]

[958 Σ—VI^h 34·6^m N 55° 52'—6, 6—5·07"—albæ, 1830, is a fine pair, which I found yellow, 1852, 1857. Omitted in the Map, but easily visible with the naked eye; the most easterly of a scattered group.]

LYRA.

For its size, one of the most remarkable constellations, and adorned by one of the great leaders of the firmament, — the first of the following list.

Double Stars.

α (Wega) to my sight is inferior to Sirius only. ♅ and H have ranked Arcturus and Capella higher: probably differences of colour affect materially the estimates which different eyes form of magnitude : * a supposition entertained by Smyth, as well as by several other great observers. Wollaston's experiments, from which he allowed Wega but ⅑ of the light of Sirius, must surely have involved some fallacy. Its colour is pale sapphire,—a lovely gem : its enormous real bulk is evident from its very minute parallax. There is a smalt-blue 11 mag. attendant at 43″ distance ; a well-known test ; my aperture of 3¹⁰⁄₁₀ inches would sometimes shew it, but could never be depended upon : I have thought it easier with 80 than 144 : it must be looked for very near the rays of *α*, as there are other minute stars at greater distances in the field.

β—xviii^h 45^m N 33° 12′.   Variable, 3 to 5 mag. in about 6^d 10½^h, two maxima and two. unequal minima occurring within that time : its three companions — 8, 8·5, 9 — with a minute neat pair, make up a fine field.   Smyth, 1834, marked *β* " very white and splendid ; " I found it, 1849, 1850, 1855, decidedly yellower than *γ*, which he calls at the same epoch " bright yellow."   H and South call it white, 1824 : Mädler pure white : Σ flava, 1835.   *γ*, on the contrary, I saw white, or very pale yellow, 1850, 1855.   Schmidt made them both whitish yellow, 1844–1855.   *γ* is suspected of variation.

ζ—xviii^h 40^m N 37° 28′—5, 5·5, 1834 [more unequal, 1850, 1855]—44″—topaz, greenish.

ε (or 4) and 5. Quadruple — xviii^h 40^m N 39° 31′—5, 6·5 and 5, 5·5—3·2″, 2·6″—yellow, ruddy ; both white.   " The naked eye," Smyth observes, " sees an irregular-looking star near Wega, which separates into two pretty wide ones under

---

* The American astronomers at Harvard College, with the great 22 feet achromatic, have found that Wega surpasses Arcturus in photographic power no less than 7 times : presumably from its different hue.

the slightest optical aid. Each of these two will be found to be a fine binary pair." So I see it, and probably most observers: there are, however, instances — Bessel at thirteen years of age was one, and I know another in England — in which the naked eye has divided ϵ and ς. There is little doubt of the rotation of each pair, ϵ perhaps in about 2,000 years, ς in half that time, and possibly both pairs round their common centre of gravity in something less than a million years. Between them lie three much smaller ones — two are excessively minute, 13 mag., on each side of the line joining ϵ and ς: in very fine weather I have had glimpses of one, and suspicions of the other: they are excellent tests for a $3\frac{7}{10}$ inch achromatic. Σ surmises alternately variable light in the components of ς. This most beautiful object, which I have seen well with $2\frac{1}{4}$ inches aperture, lies $1\frac{1}{2}° n f a$.

η—XIX^h 9^m N 38° 54'—5, 9—28"—sky-blue, violet, 1834. [yellow, greenish or bluish, in my own and Bishop's 7 inch achromatic, 1849–50. A low-power field includes two other small pairs, $s\ p$, and f.]

151 P XVIII—XVIII^h 33^m N 35° 56'—8, 9—3·8"—pale white, lilac. 3° s of a.

[δ^2 and δ^1—XVIII^h 49^m N 36° 30'—4 (on the Map, 5), 5— yellow, white. Glorious field for low powers.]

[θ—XIX^h 11^m N 37° 48'—5, 10—yellow, blue, is in a fine field.]

[About XVIII^h 30' N 35° lies a pretty pair—6, 8—yellowish, blue.]

Nebulæ.

57 M—XVIII^h 48^m N 32° 51'. The only annular nebula accessible by common telescopes; fortunately easily found midway between β and γ. It is somewhat oval, and bears

magnifying well : its light I have often imagined fluctuating and unsteady, like that of some other planetary nebulæ ; but this is of course an illusion, arising probably from an aperture too small for the object. The Earl of Rosse thought it resolvable, and saw several wisps or appendages within and without it. Secchi reduces it to minute stars, glittering like finely powdered silver. What a sight this mass of suns would be from a moderate distance ! *

56 M—XIX^h 11^m N 29° 56′. Faintish, perhaps resolvable, in a fine field and rich region, between 3° and 4° $n\ p\ \beta$ Cyg. Smyth sees it as " a globular cluster in a splendid field."

MONOCEROS.

A constellation inconspicuous to the naked eye, but rich in groups and clusters from its position in a brilliant part of the Galaxy.

Double Stars.

8—VI^h 16^m N 4° 40′—5·5, 8—13″—golden yellow, lilac.
29. Triple—VIII^h 2^m S 2° 35′—5·5, 13, 9—30″, 67″—light yellow, grey, pale blue—13 easily seen by me, 1851, 1855, 1856; more like 10 or 11 mag. : yet H and South missed it. Can it be variable ? Σ calls it 11·7 of his scale—which would be much smaller on ours—yet saw it in a 5 feet instrument.

---

* To such an object, the elegant and impressive language of Secchi may well be applied :—"Inonda il cuore un dolce senso di gioia in pensare a que' mondi senza numero, nei quali ogni stella è un sole benefico che ministro della Divina Bontà sparge vita e giocondità su altri esseri innumerabili riempiti della benedizione della mano dell' Onnipotente, e al vedersi far parte di qual privilegiato ordine di creature intelligenti, che dalla profondità dei cieli, sciolgono un inno di lode al loro fattore."

15—VI^h 33^m　N 10° 1′—6, 9·5—2·5″—greenish, pale grey. Triple in great telescopes. There are three other pairs, forming an irregular transverse line beneath it, in a glorious field.

11. Triple—VI^h 22^m　S 6° 57′—6·5, 7, 8—7·2″, 9·6″ (7 and 8, 2·8″)—white, two pale white. ♅, the discoverer in 1781, calls this " one of the most beautiful sights in the heavens." Notwithstanding appearances strongly indicative of mutual connection, no orbital motion has been detected here.

116 P VII—VII^h 21^m　S 11° 16′—7, 9·5—20″—yellow, violet.

104 P VI—VI^h 20^m　N 0° 32′—7·5, 8·5—68″—topaz yellow, plum coloured. Σ divides, Smyth elongates 8·5. A low-power field includes 77, a fine 6 mag. yellow star, with this pair *n p*, and another *s p* : a noble spectacle.

[31 Mon. and 151 P VIII—VIII^h 36^m　S 6° 35′—7, 9 on the Map. I should prefer 6, 8 : 7 must be under-rated, since it is larger than 15 Hyd. 6 mag., and is very visible to the naked eye—fine yellow, beautiful blue.]

[Region *p* 13—VI^h 24^m　N 7° 20′—very rich in wide pairs.]

Clusters.

2 ♅ VII—VI^h 23^m　N 5° 2′. Beautiful; visible to naked eye; including 12, 6 mag. yellow, and many 7 and 8 mag. stars.

50 M—VI^h 56^m　S 8° 8′. Brilliant cluster, in a glorious neighbourhood, where the Creator has

" —— sowed with stars the heaven thick as a field." *

* Milton.

27 ♅ VI—VI^h 45^m N 0° 37′. Bright Galaxy cluster, resembling three arms of a cross.

22 ♅ VI—VIII^h 7^m S 5° 23′. Group of pretty uniform 9 mag. stars.

10—VI^h 21^m S 4° 41′. This 6 mag. pale yellow star is the *lucida* of an elegant group : and the Galaxy throughout this constellation well repays the trouble of sweeping.

OPHIUCHUS.

An extensive region, somewhat barren to the eye, but attractive to the telescope.

Double Stars.

λ—XVI^h 24^m N 2° 18′—4, 6—1·1″, 1842—yellowish white, smalt blue. I unexpectedly divided this with 250, in 1854 and 1856, owing to increasing distance : a beautiful object in a large telescope. Binary : period perhaps 80 or 90 years.

67—XVII^h 54^m N 2° 56′—4, 8—55″—straw-colour, purple.

36—XVII^h 7^m S 26° 23′—4·5, 6·5—4·9″—ruddy, pale yellow, 1831, 1835, 1839, 1842 [both golden yellow, 1854, 6·5 perhaps rather deeper]. Smyth says " the principal star is thought to be variable, though I have always seen it as now registered." [1854, nearly equal, about 6·5, Smyth's smaller perhaps rather the larger.] In orbital movement, with, strange to say, a common proper motion with 30♏, more than 12′ distant, so as to lead to the impression that, in Smyth's words, " while in itself a singular revolving binary system, it

[OPHIUCHUS]

is accompanying another and a most distant object in an *annus magnus*, to contemplate the period of which makes imagination quail."

70—XVII^h 58^m N 2° 33′—4·5, 7—6·6″—pale topaz, violet. Binary, revolving in about 80 years, but with movements so singular that Jacob suspects disturbance from a third invisible companion. H observes that the rings around its telescopic image "seem to have something peculiar. They are thin, and extend further than in general ;" on another occasion he remarks, in measuring it, "Difficult, owing to the rings and appendages. N.B. I always find this star difficult from the above cause."

ρ—XVI^h 17^m S 23° 7′—5, 7·5—3·8″—pale topaz, blue. Finely grouped with two 8 mag. stars, 3° *n* a little *p* a ♏; erroneously included in ♏ in Map of S. D. U. K.

39—XVII^h 9^m S 24° 8′—5·5, 7·5—12″—pale orange, blue. Vertical near meridian. Near this beautiful pair was the Great New Star discovered by Möstlin, Kepler's scholar, in 1604; at first surpassing Jupiter and even rivalling Venus, but totally vanishing in 1¼ year, and leaving no trace whatever. 39 is easily found, 1° *n p* θ.

53—XVII^h 28^m N 9° 41′ 6, 8—41″—greyish, pale blue. 3° *s* of α.

19—XVI^h 40^m N 2° 19′—6·5, 10, 1834 [very large, 1850]—22″—pale white, livid. I found the companion large for its magnitude, 1850. Fine low-power field.

270 P XVI—XVI^h 55^m N 8° 39′—7, 8—1·5″—both white. [Test-object, elongated with 144. A line carried from ι through κ as far again will find it.]

61—XVII^h 38^m N 2° 39′—7·5, 7·5—21″ both silvery

white, 1833. [p considerably larger than f, 1850. Fletcher made them 6, 7, both yellow, 1851. Closely p γ, and 2° nearly s from β, a fine specimen of a pale yellow star.]

[2191 Σ—XVII^h 30·5^m S 4° 51—7, 7·5—both white, 2° s f 14 M.]

[21—XVI^h 43^m N 1° 30′—6 mag. yellow, has a wide 8, 9 mag. pair in field.]

(Temporary—XVI^h 51^m S 12° 39′, about 3° np η. Hind's New Star, 5 mag. 1848, April 28,* which subsequently faded to a minute point, and may yet blaze out again : it was orange, with flashes of red, a colour which I have seen in a few other stars.)

Clusters and Nebulæ.

23 M—XVII^h 49^m S 18° 59′. Glorious low-power field.

10 M—XVI^h 50^m S 3° 54′. Bright resolvable nebula. There is a beautiful group f; the principal star bright orange.

12 M—XVI^h 40^m S 1° 43′. Resolvable; finely grouped.

19 M—XVI^h 54^m S 26° 4′. Large, and tolerably bright, but very near horizon.

14 M—XVII^h 30^m S 3° 10′. Large, with glimpses of resolution, which was effected by ♓ with the 20 feet reflector.

9 M—XVII^h 11^m S 18° 22′. Small, apparently resolvable : "a myriad of minute stars, clustering into a blaze in the centre, and wonderfully aggregated." (Smyth.)

[About XVII^h 38^m N 5° 40′, closely n f β, is a bright group, in appearance evidently a *family*: among them is a pair—7, 9—pale blue, blue.]

* Airy considered it fully 5 mag. May 9, and 4 mag. May 10, sinking to 6 mag. May 18.

ORION.

The finest constellation in the heavens, equally remarkable
for telescopic interest and obvious brilliancy; fortunately its
position is very suitable for an observer in our latitude, as it
comes to the meridian in winter, and does not attain an incon-
venient altitude.

α (Betelgeuze) is irregularly variable. H, the discoverer,
found it alternately above β (Rigel), and below α ♉ : from
1839, Nov. 26 to 1840, Jan. 7, losing nearly half its light:
afterwards its changes were much less conspicuous. 1852,
about Dec. 5, Fletcher thought it brighter than Capella,
which he rated clearly above Wega, so that it was then the
brightest star in the N. hemisphere. Pogson finds its maxima
in April and October. He ranks it above β; his assistant,
Quirling, the reverse, as I should do. Lassell says of it,
"a most beautiful and brilliant gem ! Singularly beautiful
in colour, a rich topaz; in hue and brilliancy different from
any other star I have seen." Look at α and β alternately;
even a small telescope will shew the beauty of the contrast.

Double Stars.

β—v^h 8^m S 8° 22′—1, 9—9·5″—pale yellow, sapphire
blue. I see always a blue tinge in the great star, resembling
that of Wega. Σ however gives albasubflava; Fletcher
yellow. A beautiful object, and a fair test for a pretty good
telescope; but from its low altitude often blotted with
vapour.

δ—v^h 25^m S 0° 24′—2, 7—53″—pale white, flushed

white, 1850 [2 beautiful pale green, 1850; white or pale yellow, 1851]. This star nearly marks the equator.

ζ. Triple—v^h 34^m S 2° 1'—very large 3, 6·5, 10—2·5", 56"—topaz yellow, light purple, grey. 6·5, singularly missed by ♅, and discovered by Kunowsky, seems of some nondescript hue, about which observers cannot agree. Σ calls it in one specially manufactured word, "olivaceasubrubicunda."*

Triple—v^h 29^m S 6° 0'—3·5, 8·5, 11—12", 50"—white, pale blue, grape red. Field very fine: I think I have seen a glow around this group, which the Earl of Rosse finds to occupy a singular dark opening encompassed by nebulous matter.

λ—v^h 27^m N 9° 50'—4, 6—4·5"—pale white, violet, 1833, 1843. Dembowski yellow, blue, "couleurs sûres," 1856. A very fine field.

σ. Multiple—v^h 32^m S 2° 41'—4, 8, 7—13", 42"—bright white, bluish, grape red, 1832, 1850 [4 yellow, 1851]. There is an 11 mag. star in the group, so small as to have escaped some of the first observers. It is preceded by a beautiful little triangle p, 8·5, 9, 8—dusky, white, pale grey.

32—v^h 23^m N 5ᵛ 50'—5, 7—1"—bright and pale white. Fine test. I could only elongate it suspiciously. It lies f γ, a little s.

23—v^h 15^m N 3° 24'—5, 7—32"—white, pale grey. 3° s of γ, a little p.

ρ¹—v^h 6^m N 2° 41'—5, 8·5, 1835 [8·5 very small 1849, 1856]—6·8"—orange, smalt blue.

* This almost matches Gruithuisen's *mondfangsphärenhalbmesser*, and *stickstoffsauerstoffatmosphäre*, or Von Boguslawski's *sternschnuppen-beobachtungen*. An English chemist speaks of *iodide of methylodiethyla-mylammonium*.

[ORION]

52—v^h 40^m N 6° 24'—6, 6·5—1·8"—pale white, yellowish.
Σ reverses the colours. I see discs in contact, 80, neatly
divided, 144: an excellent test, readily found, about 2° *s p a*.

33—v^h 24^m N 3^v 11'—6, 8—2"—white, pale blue. Between γ and ζ, nearer γ.

278 and 279 P IV—iv^h 55^m N 1° 24'—8·5, 9—14"—
silvery white, pale blue. [1° *f* π⁶, 5 mag. About 1° *f*
this pair, a little *s*, is an 8 mag. star, of a singularly deep and
striking red hue.]

258 P IV—iv^h 51^m N 1° 27'—8·5, 9—2·4"—white, pale
grey. This beautiful object may be binary. It is closely
s p π⁶.

[859 Σ—vi^h 0^m N 5° 40', and 809 Σ—v^h 42^m S 1° 28',
are pairs worth looking for.]

Clusters and Nebula.

362 H—v^h 29^m S 4° 27'. Brilliant field, containing pair—
6, 9—5"—lucid white, pale blue. In a grand neighbourhood:
in fact the whole space should be well swept over from
42 to *ι*.

24 ♅ VIII—vi^h 1^m N 13° 59'. Triangular cluster containing pair—7·5, 8·5—2·4"—both lucid white. 1° *s* of *ν*.
A remark of Smyth's in this place must be transferred to these
pages. "These gatherings, occurring indifferently upon the
Via Lactea and off it, awaken still more our admiration of the
stupendous richness of the Universe, in every department of
which there appears such a profusion of creation, if we may so
express ourselves of the works of the ALMIGHTY, in which our
utmost ken has yet never detected any redundancy, much
less anything made in vain."

θ^1—v^h 28^m S $5°$ $29'$, points out the GREAT NEBULA IN ORION, one of the two most wonderful objects of this character in the heavens; so distant that, till very lately, its starry nature was doubted; so dense as to be visible to the naked eye : * that of Andromeda alone stands in the same class : other nebulæ, if equally bright, are easily resolved; if irre-solvable, are obscure : in these objects there must be either individual splendour, or general condensation, elsewhere unrivalled. The one before us will be found an irregular branching mass of haze, in some directions pretty well de-fined where the dark sky penetrates it in deep openings, analog-ous, perhaps, to the dark streaks in the nebula in Andromeda; in others melting imperceptibly away over such an extent that Secchi, by moving the telescope rapidly to gain full contrast, has traced it in singular convolutions, and with a consider-able break near σ, through $5\frac{1}{2}°$ of Dec. and $4°$ of R A,—from ζ to 49, and probably to 38 ♅ V—a prodigious diffusion. Many stars are visible throughout it, but all far on this side; two or three instruments only in the world have yet approached to an insulation of the inexpressibly minute points which form the haze. It resisted ♅'s 40 feet reflector, in which it was the first object viewed, and, together with that in Andromeda, suggested to him the widely-discussed Nebular Hypothesis, which would see here nothing more than an unformed fiery mist, the chaotic material of future suns. H found but the aspect of " a curdling liquid, or a surface strewed over with flocks of wool, or the breaking up of a mackarel sky." The

---

* It was strangely missed, as Humboldt says, by Galileo, who paid great attention to Orion. Wolf finds that Cysatus saw it with the telescope previous to 1619.

[ORION]

Earl of Rosse, with his admirable 3 feet reflector, Lassell, with his perfect 2 feet speculum in the Maltese sky, could advance no further : it was reserved for the 6 feet mirror of the Earl of Rosse, as the reward of his munificent perseverance, to lift the veil, and distinguish in some places its starry composition. Bond independently arrived at the same conclusion; and Secchi, with smaller, but very perfect means, detects the glittering "star-dust," which seems certain of future and more complete developement. Yet, though this would imply a permanent form, there are strange discrepancies in the draw-ings by the best hands. H in England, the same observer at the Cape of Good Hope, Bond, Lassell, Liapounov with a $9\frac{1}{2}$ inch achromatic at Kazan, Otto Struve (jun.) at Poulkova, all differ in various ways; the latter even believes that the bright-ness of the central part is in a state of continual change : either then we have before us a singular agreement in a delu-sion, or an inexplicable wonder. Any confident opinion would seem to be premature, and here may be secrets laid up for future generations.* In the densest part, four stars, — 6, 7, 7·5 and 8 mag., — of which θ^1 is the leader, — form a trapezium. Smyth gives their colours pale white, faint lilac, garnet, reddish; and Mädler thinks their connection is shewn by a common movement in space. A powerful telescope adds a fifth star, which is believed to have become visible only of late years. Dawes has seen it with a 5 feet achromatic, but it is no disparagement to other observers to fail where he

---

* Rerum natura sacra sua non simul tradit. Initiatos nos credimus; in vestibulo ejus hæremus. Illa arcana non promiscue nec omnibus patent, reducta et in interiore sacrario clausa sunt: ex quibus aliud hæc ætas, aliud quæ post nos subibit, dispiciet. Tarde magna pro-veniunt.—SENECA, *quoted by Humboldt.*

succeeds.   I have suspected it, but very uncertainly.   A sixth still smaller has since been perceived, and two or three other most minute points; one has been added by Porro, who has been for some time preparing at Paris an enormous object-glass of $20\frac{1}{2}$ English inches in diameter.   Otto Struve is disposed to think that several of them are subject to change, and remarks that "the existence of so many variable stars on such a small space in the central part of the most curious nebula in the heavens must of course induce us to suppose these phænomena intimately connected with the mysterious nature of that body."   The observer will notice how beautifully one considerable star, nearly opposite the great dark opening, is encompassed by a separate cloud of mist.   The clearest weather must of course be chosen, and the lowest power which will bring out the trapezium is most likely to give a satisfactory contrast with the exterior darkness.

PEGASUS.

A constellation easily recognized by the great square which three of its principal stars form with that in the head of Andromeda.

Double Stars.

ε — xxiʰ 37ᵐ   N 9° 14′ — 2·5, 9 — 138″ — yellow, violet. This object, when near the meridian, well exhibits a phænomenon noticed by H,— the apparent oscillation, as of a pendulum, of a small star in the same vertical with a large one, when the telescope is swung from side to side; this he thinks is due to the longer time required for a fainter light to affect the retina, so that the reversal of motion is first perceived in the brightest object.

1 — xxi^h 16^m N 19° 12'—4, 9 — 36"— pale orange, purplish. 4 said to be variable, and common proper motion suspected.

κ — xxi^h 38^m N 25° 0'— 4, 13, 1836 [more like 11, 1852] — 12"—pale white, purple. A good test for light.

π^1 — xxii^h 3^m N 32° 30'—5—yellow, forms a grand pair with π^2, 4 mag. [yellow.]

3— xxi^h 31^m N 5° 59'—6, 8—39"—flushed white, greyish : another very pretty pair in field, 8" apart.

33 — xxii^h 17^m N 20° 9'—6·5, 8 —58"— yellowish, pale grey. Σ gives them common motion through space. 6·5 has a 10 mag. attendant at 2·7", which I could not see. This object is the leader of a line of 6 similar stars.

306 P XXII—xxiii^h 1^m N 32° 4'—7, 8·5—8·5"— bright white, sapphire blue. [Several little pairs similar to each other lie dispersed in this region.]

216 and 217 P XXIII — xxiii^h 46^m N 11° 9'—8·5, 8·5—18"—both silvery white, 1834 ; so Σ 1830, yet I saw them white, pale blue, 1850–1, not quite alike, 1856. The mags. I also found obviously unequal, p the smaller : I was not then aware that Σ had found the difference vary a whole mag. The period of this change should be investigated.

[η — xxii^h 35^m N 29° 25'— has a bluish 10 mag. companion, resembling ε, but the large star is of a paler yellow. Schmidt thinks its tint variable—more or less white or red in different years.]

(29 of Σ's nomenclature — xxi^h 24·9^m N 19° 58'—7, 7·5— 2·7"—albæ, must be a neat pair. I have not seen it.)

Nebula.

15 M—XXI^h 23^m N 11° 33'. Bright and resolvable, blazing to centre : a very fine specimen of a completely insulated cluster.   Discovered by Maraldi, 1745.

PERSEUS.

Here again we enter upon one of the most splendid portions of the Galaxy.   Night after night the telescope might be employed in sweeping over its magnificent crowds of stars.   This constellation includes the most conspicuous of, at least, the regularly variable stars, β, or Algol (11^h 59^m N 40° 25') which changes from 2 to 4 mag. in a few seconds less than 2^d 20^h 49^m, the increase and decrease together occupying not more than 7^h, the minimum only 18^m : so that it usually appears 2 mag. There are, however, slight irregularities in these times,—probably the slow working of some unknown general law, which may affect all these wonderful bodies.

Double Stars.

ε — III^h 48^m N 39° 36'—3·5, 9, 1832—8·4''—pale white, lilac. [9 very small for this mag. 1849, on several occasions, once with Bishop's achromatic, 7 inches aperture, in the Regent's Park.]

ζ.   Quadruple — III^h 45^m N 31° 28'—3·5, 10, 12, 11 (1832) —13'', 83'', 121''—flushed white, smalt blue, ashy, blue.   The attendants seemed to me, 1850, to increase their size with their distance, the closest being much more minute than the others. ♄ also appears to have seen but three ; can one be variable ?

η—II^h 40^m N 55° 19′—5, 8·5—28″—orange, smalt blue.

220 and 222 P II—II^h 51^m N 51° 48′—6, 8—13″—silvery white, sapphire blue. Visible to naked eye, forming a triangle with γ and τ.

20—II^h 45^m N 37° 46′—6·5, 10—14″—pale white, sky blue. "A neat test object," (Smyth) which my $3\frac{7}{10}$ inch aperture shewed readily and perfectly. Closely f 16, 5 mag.

12—II^h 33^m N 39° 36′—6—yellow, has two pairs near it, in a large field.

58—IV^h 27^m N 40° 58′—5·5—orange, has a pair in field —7·5, 9—12″—greenish, lilac.

57—IV^h 24^m N 42° 46′—8, 8—110″—both white, 1833 [not exactly alike in colour, 1852].

80 ♅ VIII—III^h 39^m N 52° 14′—8, 11—9·5″—light yellow, pale violet.

[37 P III—6 mag.—orange—near a, s a little p, has a fine blue companion in a beautiful field.]

[425 Σ—III^h 29^m N 33° 34′—8, 8—2·9″—both white. A true "pair," a little p 40, 6 mag.]

[443 Σ—III^h 35^m N 40° 56′—8, 9—9″—both white. The second star nearly s of ν_r about 1° distant.]

Clusters.

33 and 34 ♅ VI—II^h 9^m N 56° 30′. These two gorgeous clusters, described by Smyth as "affording together one of the most brilliant telescopic objects in the heavens," are visible to the naked eye as a protuberant part of the Galaxy, and so ♅ considered them. They are often called *the sword-hand of Perseus*. With 64 these superb masses were visible together, as well as a brilliant pair n.

34 M—11ʰ 33ᵐ N 42° 8'. Just perceptible without the telescope; a very grand low-power field, one of the finest objects of its class. It contains a little 8 mag. pair, 14″ apart.

227 H—11ʰ 23ᵐ N 56° 54'. Wide cluster, a little *f* the sword-hand.

25 ♅ VI—111ʰ 5ᵐ N 46° 43'. A low power shews a very faint large cloud of minute stars, beautifully bordered by a foreshortened pentagon of larger ones.

PISCES.

A dull region, containing some good telescopic objects.

Double Stars.

α—1ʰ 55ᵐ N 2° 5'—5, 6—3·8″—pale green, blue, 1834, 1838, 1850; Σ ditto, 1831; Fletcher both yellow, 1851; Dembowsky white, ashy white, 1854. The contrast is certain, but I found 6 troublesome as to colour, usually ruddy or tawny, sometimes bluish; see the remark under 69 P XIV Boöt., p. 174. 1855, I noted it pale yellow, brown yellow; " quite satisfactory." Fletcher and Jacob think this fine object may be binary.

ψ^1 and ψ^2—0ʰ 58ᵐ N 20° 43'—5·5, 5·5—30″—flushed white, pale white.*

65 — 0ʰ 42ᵐ N 26° 57'—6, 7—4·5″— both pale yellow. [Very little, if at all, unequal, 1850, *p* possibly the smaller: the same, by an independent estimate, 1855. I have since met with Σ's precise concurrence, " æquales—fortasse præ-

---

* It should have been stated in the proper place, p. 159, that a few of these colours (as in this instance) have been taken from Smyth's subsequent revision in the " Ædes Hartwellianæ."

cedens paululo minor "—so that there seems a fair case of variation, especially as Smyth observes that P had made Smyth's 6 his own companion star, and therefore seen it of lesser magnitude.]

35 — 0^h 8^m N $8°$ $3'$—6, 8—$12''$— pale white, violet.

ζ — 1^h 7^m N $6°$ $50'$—6, 8—$23''$—silver white, pale grey. 6 has been suspected variable up to 4. Σ gives common proper motion.

55 — 0^h 33^m N $20°$ $40'$—6, 9—$5\cdot9''$— orange, deep blue, " a rich specimen of opposed hues," 1833. Dembowski yellow, deep red, "couleurs sûres," 1856. [9 very small, and its colour indistinct, 1848, 1850.]

123 P I—1^h 29^m N $6°$ $56'$—$6\cdot5$, 8—$1\cdot4''$—yellowish, pale white. Binary ? A fine test, requiring beautiful weather: elongated, 80; in contact, 144; separated, 250. It lies between μ and o, but there are several similar stars around: look for a long, narrow trapezium in the finder; it will be the $s\,p$ of the four stars: in the telescope it has a 10 mag. star $s\,f$ at a little distance.

51—0^h 25^m N $6°$ $11'$—$6\cdot5$, 9—$28''$—pearl white, lilac.

100 — 1^h 27^m N $11°$ $51'$—7, 8 — $16''$—white, pale grey. Binary ? Closely $n\,p\,\pi$.

38—0^h 10^m N $8°$ $6'$—$7\cdot5$, 8—$4\cdot8''$—light yellow, flushed white. Σ common proper motion. Closely f 35, *supra*.

77—0^h 59^m N $4°$ $10'$—$7\cdot5$, 8—$32''$—white, pale lilac.

251 P O—0^h 52^m N $0°$ $2'$—8, 9—$18''$—pale orange, clear blue. Probably binary.

[155 Σ—1^h 35^m N $8°$ $34'$—$7\cdot5$, $7\cdot9$—$4\cdot6''$—both white. Beautiful field, a little $n\,p\,o$.]

[85 and 87 P I—1^h 20^m N $7°$—7, 9—yellow, bluish.]

SAGITTA.

A little asterism, of much greater antiquity than might have been supposed from its size and the comparative smallness of its components.

Double Stars.

ζ—xix^h 43^m N 18° 48'—5, 9—8·6"—silvery white, blue, 1831, 1838. Dembowski 5 yellow, 1856.

ε—xix^h 31^m N 16° 9'—6, 8—92"—faint yellow, bluish.

θ. Triple — xx^h 4^m N 20° 30'— 7, 9, 8 — 11", 70"— pale topaz, grey, pearly yellow.

[10 and 11, two 6 mag. stars, form an object worth looking for, in a rich field.]

[13, 6 mag., orange, is the *lucida* of a beautiful group, of which the nearest is a pretty little 10 mag. pair.]

[15, 6 mag., commands another fine group. n a little p, at a few minutes' distance, is 392 P XIX, a beautiful sapphire blue 7 mag. star.]

[η, 6 mag., yellow, lies in a rich region. A circle around it of 30' or 40' radius will include several very pretty little 8 or 9 mag. pairs, on different sides of it.]

Cluster.

71 M—xix^h 47^m N 18° 25'.   Large and faint, hazy to low powers with $3\frac{7}{10}$ inches aperture, yielding a cloud of faint stars to higher magnifiers; an interesting specimen of the process of nebular resolution. It lies in the Galaxy, rather more than 1° $s\,p\,\gamma$. About 1° $s\,p$ 71 M is a beautiful low-power field, containing a pair and a triple group, all about 8 or 9 mag.

SAGITTARIUS.

The stars of this constellation have a beautiful effect above the S. horizon, near the place where the Galaxy passes from our sight; but in our latitude they are apt to be obscured by vapour. The Milky Way is here very rich in a sufficiently transparent night.

Double Stars.

μ^1. Triple—xvIII^h 5^m S 21° 56'—3·5, 9·5, 10—40", 45" —pale yellow, two reddish, 1835. [10 bluish, equal at least to 9·5, 1850, 1855.]

54 — xIx^h 33^m S 16° 37'—5·5, 8—28"—yellow, violet. Fine field of minute stars.

[About 1¼° s of λ, xvIII^h 18^m S 25° 30' is a fine 7 mag. triangle, the s and f stars of which have smaller attendants.]

Clusters and Nebulæ.

22 M—xvIII^h 28^m S 24° 1'. Beautiful bright cluster; very interesting from the visibility of the components, the largest 10 and 11 mag., which makes it a valuable object for common telescopes, and a clue to the structure of more distant or difficult nebulæ. It lies midway between μ and σ.

25 M—xvIII^h 23^m S 19° 10'. Coarse and brilliant.

21 M—xvII^h 56^m S 22° 31'. In a lucid region.

30 ♉ VII—xvIII^h 4^m S 21° 36'. Curious large undefined cloud of 10 mag. stars; requiring low power and steady gazing; ½° s of μ^1. Erroneously marked 30, II, on Map.

28 M—xvIII^h 16^m S 24° 56'. Not bright: seems resolvable. 1° $n\,p\,\lambda$.

75 M — xIx^h 58^m S 22° 19'. Bright nucleus with low power.

[SAGITTARIUS]

51 ♄ IV — XIXʰ 36ᵐ S 14° 29'. Planetary: like a star out of focus. Secchi thinks it resolvable. 2ᵛ *n*, a little *f* 54, *antea*.

[8 M — XVIIʰ 53ᵐ S 24° 50'. Splendid Galaxy object ; visible to naked eye. In a large field we find a bright coarse triple star, followed by a resolvable luminous mass, including two stars, or starry centres, and then by a loose bright cluster enclosed by several stars : a very fine combination.*]

SCORPIO.

A fine constellation, little noticed by casual star-gazers, from its low altitude and short continuance above the horizon, with the additional disadvantage of its reaching the meridian during the brief summer's night. The student will do well

---

* A little *n p* μ—XVIIʰ 57ᵐ S 18° 50'—is a spot referred to by Secchi as exemplifying in a high degree the marvellous structure which he has observed in the Galaxy, with the great achromatic at Rome. The remarks of this accomplished astronomer on the successive layers of stars are very curious: first he finds large stars and lucid clusters; then a layer of smaller stars, certainly below 12 mag.; then a nebulous stratum with occasional openings. But what he says startled him, and all to whom he shewed it, was the regular disposition of the larger stars in figures "si géométriques qu'il est impossible de les croire accidentelles. La plus grande partie sont comme des arcs de spirale ; on peut compter jusqu'à 10 où 12 étoiles de la 9ième à la 10ième grandeur . . . . se suivant sur une même courbe comme les grains de chapelet; quelquefois elles forment des rayons qui semblent diverger d'un centre commun, et ce qui est bien singulier, on voit d'ordinaire que, soit au centre des rayons, soit au commencement de la branche de la courbe, on trouve une étoile plus grande et rouge. Il est impossible de croire que telle distribution soit accidentelle." He mentions, besides this spot, several instances in Cygnus. What a remarkable parallel to the spiral structure discovered by the Earl of Rosse in so many nebulæ! See also Smyth's remark on 35 M II, p. 199, *antea*.

to look out for it, and it will repay an hour or two of extra watching.

Double Stars.

α (Antares)—XVI^h 21^m S 26° 7′. This great star Smyth justly terms "fiery red :" and it is a grand object in the telescope. Its tint, however, is not uniform : to me the disc appears yellow, with flashes of deep crimson alternating with a less proportion of fine green.   This latter mixture has perhaps been subsequently accounted for by the discovery (in 1846, by Mitchell in America) of a 7 or 8 mag. green star about 3″ from the principal, and therefore involved in its flaming rays : Secchi thinks it may be variable, as it has not always been seen.   Dawes noticed a curious proof of its independent, not contrasted green light, when it emerged, in 1856, from behind the dark limb of the Moon before its overpowering neighbour.

β—XV^h 58^m S 19° 25′—2, 5·5—13″—pale or yellowish white, lilac.

ν—XVI^h 4^m S 19° 6′—4, 7—40″—pale yellow, dusky. Jacob subdivides 7 into 7, 8—1·75″; and so I see it with my present very fine 5½ inch object-glass by Alvan Clark.   It might doubtless be perceived with smaller telescopes, and could not have escaped Smyth in 1831, had it then been as conspicuous as it is now.

σ—XVI^h 13^m S 25° 15′—4, 9·5—20″—creamy white, lilac.

236 P XVI—XVI^h 49^m S 19° 19′—6·5, 8—5·8″—yellowish white, pale green. Binary ?

31—XVII^h 9^m S 26° 28′—6·5, 11—6·8″—pale white, ash-coloured.   Identical with 38 Oph.

[SCORPIO]

48 and 49 P XVI — xvi^h 12^m S 19° 47′ — 8, 9 — 14″ —
dull white, flushed.   Another pair in field, forming a beautiful group.   1° p ψ Oph.

[ρ — xvi^h 16^m S 23° 0′ — 6, 7 — a beautiful close pair, has
two other 7 mag. stars in the field, 68 and 72 P XVI.]

Nebulæ.

80 M — xvi^h 9^m S 22° 39′.   Like a comet; in a beautiful
field, half way between α and β.   ♓ calls it the richest and
most condensed mass of stars in the firmament.

4 M — xvi^h 15^m S 26° 10′.   Large, rather dim, resolvable,
followed by a vacant space without stars distinguishable in
my telescope.   1$\frac{1}{2}$° p a.

SERPENS.

A long, rambling constellation, mixed with Ophiuchus.   It
contains some fine telescopic objects.

Double Stars.

δ — xv^h 28^m N 11° 1′ — 3, 5 — 2·8″ — both bluish white,
1831 — 1842.   Dembowski yellow, ashy yellow, 1853 — 1855;
whitish yellow, ashy olive, 1856.   Fine specimen of the class
of moderately close unequal pairs.   Possibly binary.

θ¹, θ² — xviii^h 49^m N 4° 1′ — 4·5, 5 — 22″ — pale yellow,
golden yellow.   Smyth says 4·5 has been variously rated, and
should be watched for variable light: this would be easy, close
to so excellent a standard of comparison; but to prevent mistakes, it must be borne in mind that θ¹, or 4·5, was p, 1834.
Σ finds common proper motion.   This noble pair lies in a dark
space between the two streams of the Galaxy.   N.B. There is

a traditional misrepresentation of the latter in this region; where I have found both streams usually misplaced on globes and maps; p being at once too narrow and too far W.: the centre of this branch, marked by 72 ♅ VIII, is really midway between 72 Taur. Pon. and θ Serp. on either side of it. H has given in his " Outlines of Astronomy " a very accurate description of the Galaxy, which shews how little the common representation of it is to be trusted.

59 — xviii^h 20^m N 0° 7′—5·5, 8—3·9″— yellow, indigo blue.   Brightest of the vicinity.

49 — xvi^h 7^m N 13° 54′—7, 7·5—3·3″— pale white, yellowish.   Binary in 600 years?

220 P XV — xv^h 50^m N 3° 49′—8, 9—11″—white, grey, 1834. Dembowski 8 blue, "sûre," 1856.   Binary? $1\frac{1}{2}°$ sf ε.

[2017 Σ — xvi^h 4·2^m N 14° 58′—7·7, 8·4—25″—yellow, white.   A pretty pair.]

Clusters.

[72 ♅ VIII — xviii^h 20^m N 6° 30′.   Very fine, with a 6 mag. star in the field : visible to naked eye.   Between it and θ, nearer the former, is a beautiful large cloud of stars, chiefly 8 and 9 mag., a nearer part, apparently, of the Galaxy : visible to naked eye, and requiring a large field.]

SEXTANS.

A modern asterism, as its name denotes, being one of the minor constellations formed by Hevelius out of unclaimed stars lying between the ancient ones.

35 — x^h 36^m N 5° 29′—7, 8—6·8″—yellow, blue, 1834, 1839, 1849. Σ ditto, 1832. [yellow, ruddy, 1852.]   In a fine field.

9—IX^h 47^m N 5° 36′—7, 9—50″—flushed blue, pale blue, 1851 [red, blue, 1852]. Pointed at by α and π Ω.

Nebulæ.

163 ♅ I—IX^h 58^m S 7° 3′. Very distinct, with a much brighter centre, bearing magnifying unusually well.

4 ♅ I — X^h 7^m N 4° 9′. Two very faint nebulæ, in a glorious field. ♅ missed the fainter of these, though he observed the other four times.

TAURUS.

An interesting constellation, containing two beautiful groups familiar to the first beginner in stellar astronomy,—the Pleiades, and Hyades. Neither of these, however, is sufficiently concentrated to make a good telescopic object, excepting in an unusually large field. The 6 principal stars of the Pleiades are evident to any clear sight; but glimpses of more are easily attainable. Möstlin is said by Kepler to have distinctly made out 14. A beautiful triangle of small stars will be found near the *lucida*, Alcyone.

Double Stars.

α — IV^h 28^m N 16° 14′—1, 12 — pale rose-tint [yellow]. Al-debaran, in Arabic *the hindmost*, because he seems to drive the Pleiades before him. The extremely minute attendant is a good light-test. Dawes has seen it with a 2$\frac{3}{4}$ inch glass: mine of 3$\frac{7}{10}$ inches shewed it certainly, but not without much attention: 144 suited it better than 80: it lies at some distance *n f*, rather more *n* than *f*. Occultations of Aldebaran are not infrequent, as it lies in the Moon's way: they are worth looking for, as they often exhibit the very strange and unexplained phænomenon of the projection of the star upon the Moon's disc.

θ^1 and θ^2—IV^h 21^m N 15° 39'—5, 5·5—337"—pearly white, yellowish. Common proper motion? Among Hyades; visible to naked eye as a pair.

τ—IV^h 34^m N 22^u 41'—5, 8—62"—bluish white, lilac.

88—IV^h 28^m N 9° 52'—5, 8·5—68"—bluish white, cerulean blue. Binary?

χ—IV^h 14^m N 25° 18'—6, 8—19"—white, pale sky blue or grey.

80—IV^h 22^m N 15° 20'—6, 8·5—1·8", 1843—yellow, dusky. I could only see it single, 1851, but being probably binary, it may open out so as to be more accessible. It lies in the Hyades, about $1\frac{1}{2}°$ $s\,p\,a$.

ϕ—IV^h 12^m N 27° 1'—6, 8·5—56"—light red, cerulean blue.

111—V^h 16^m N 17° 15—6, 8·5—63"—white, lilac, 1832 [yellow, lilac, 1851].

30—III^h 41^m N 10° 43'—6, 10—9"—pale emerald, purple. [10 a mere point: better with 80 than 144.]

118—V^h 21^m N 25° 2'—7, 7·5—5"—white, pale blue. Between tips of horns, nearer n.

257 P IV. Triple—IV^h 51^m N 14° 20'—7, 8, 10—39", 70"—white, cerulean blue, purple. $n\,f\,o^1$ and o^2 Orion.

62—IV^h 16^m N 23^u 58'—7, 8·5—29"—silver white, purple. In a fine field.

37 P V—V^h 11^m N 19° 59'—7, 11—9"—deep yellow, bluish. The last of a curious series of 6 stars nearly following each other.

213 P III. Triple—III^h 53^m N 22° 48'—7·5, 8, 12 (1835)
—7·5", 60"—white, grey, blue. [8 more like 9 : 12 like 10,
1855.] 12 is probably variable. South missed it, 1823. Σ
gave it different rates in different years. 3° f Pleiades, a
little s.

20 P V—V^h 9^m N 18° 17'—8, 8·5—2·1"—both bluish.
144 divided it neatly.

[σ^1 and σ^2—IV^h 30^m N 15° 30'—5, 5—both pure white,
look like a connected system : so do κ^1 and κ^2, placed by Σ in
IV^h 15·6^m N 21^v 52'—5, 6.]

[559 Σ—IV^h 23·5^m N 17^v 38'—7, 7—3"—both white.
Between α and ε, rather nearer α, f the line joining
them.]

Nebula.

1 M—V^h 25^m N 21° 55'. Oblong; pale; 1° np ζ, on
s horn. The Crab Nebula of the Earl of Rosse, who resolved
it, as well as brought out its curious fringes and appendages.
Secchi has obtained the same result. Its accidental discovery
by M, while following a comet in 1758, led to the formation
of the earliest catalogue of nebulæ.

TAURUS PONIATOWSKII (or PONIATOVII.)

A little asterism in a very rich and beautiful part of the
Galaxy.

Double Stars.

362 P XVII—XVII^h 59^m N 12° 0'—8, 8·5—6·9"—straw
yellow, sapphire blue, 1831, 1838. Dembowski white,
green, 1855–6. I thought the magnitudes underrated, 1850.

[TAURUS PONIATOWSKII]

[About xvii^{h} 55^{m} N 7° 45′ is a wide 8, 8·5 mag. pair. It is 1¼° $s\,p$ 71, 6 mag.]

(78—according to Σ, xviii^{h} 42·6^{m} N 10° 47′—6, 7·5— 3·5″—yellow, blue, " colores insignes," must be a fine object: I have not looked for it.)

TRIANGULUM.

One of the ancient constellations, containing several good objects.

Double Stars.

ι—ii^{h} 4^{m} N 29° 39′—5·5, 7—3·5″—topaz yellow, green : " exquisite," Smyth. Marked only 6 in Map.

160 P II—ii^{h} 36^{m} N 28° 52′—8, 8·5—2·9″—both cream-white, 1831 [white, yellowish or ruddy, 1850].

38 and 39 P II—ii^{h} 9^{m} N 28° 6′—8·5, 9—14″—both silvery white, 1834. [yellowish, greyish or bluish grey, 1849, 1852, with more than ½ mag. difference. Closely p 10, 6 mag.]

[197 Σ—i^{h} 50·8^{m} N 34° 27′—8, 9 — 18″. A pretty pair, 2° $n\,p\,\beta$.]

[232 Σ — ii^{h} 4·6^{m} N 29° 35′— 6·6″ — 8·5, 8·5 — both white. Closely $f\,\iota$.]

Nebula.

33 M—i^{h} 26^{m} N 29° 58′. Very large, faint, ill-defined ; visible from its great size in finder, a very curious object, only fit for low powers, being actually imperceptible, from want of contrast, with my 144. It was resolved by ♅ into stars " the smallest points imaginable," in which the Earl of Rosse finds the same spiral arrangement which prevails so wonderfully in many nebulæ.

URSA MAJOR.

This familiar constellation offers a large field to the perse-vering observer. It must be borne in mind that it extends far beyond the region occupied by "the seven stars"; and from the unmarked character of some parts of it, several telescopic objects will require some pains in their identification. It seems difficult to ascertain whence this Bear and his com-panion derived their preposterous length of tail. Dr. Mather, in 1712, tells a curious story, that though the Indians did not divide the stars into constellations, they called the stars of Ursa Major, *Paukunawaw*, that is, the Bear, long before they had any communication with Europeans.

Double Stars.

α—x^h 55^m N 62° 30'—1·5, 8—381"—both yellow, 1832 [8 violet, 1850]. H suspects 1·5 of variable light.

ζ—xiii^h 18^m N 55° 39'—3, 5—14"—brilliant white, pale emerald. This fine pair, which is said to have been discovered 1700, Sept. 7, by Godfrey Kirch and his scientific wife Mary Margaret, and which may possibly be travelling together through space, forms a noble group with Alcor, 5 mag., 11½' distant (the " rider upon the horse,") and another smaller star. ζ, or Mizar, and Alcor, form a pair to the naked eye: and thus become an excellent object for a beginner, as the tele-scopic increase of brightness and distance admits of direct comparison ; but the inversion of the astronomical eye-piece must be borne in mind, or the identity will be perplexing. There is a strange story as to the large star's having been found single by Mädler, at Dorpat, 1841, Apr. 18, before sun-

set, and in twilight, though more difficult pairs were perfectly
well seen: within 1^h afterwards it had become double as
usual.   It is strange that he did not watch it closely enough to
catch it in the very act of reappearing.

ξ—xi^h 11^m N 32° $19'$—4, $5\cdot5$—$2\cdot3''$—subdued white,
greyish white.   Binary, in 59 years, H; 65 years, Smyth.
It has been watched round more than a complete revolution.

23—ix^h 20^m N 63° $40'$—4, $9\cdot5$—$23''$—pale white, grey.

σ^2—viii^h 58^m N 67° $42'$—$5\cdot5$, $9\cdot5$—$5''$—flushed white,
sapphire blue.   Binary?   If so, we have here again a striking
proof that the smaller stars are not invariably most distant
from the earth.   The attendant, very difficult with 80, was
plainly seen with 144.

111 P XI—xi^h 29^m N 28° $33'$—6, 7—$1\cdot4''$—both pale
blue. [elongated only with 250.]   Smyth says, " it is situated
in a very vacant space to the eye . . . but to the powerful
reflectors now in use, is in a very ocean of nebulæ."   I missed
a 13 mag. star which makes it triple.

156 P XIII—xiii^h 32^m N 51° $26'$—6, 8—$1\cdot9''$—topaz
yellow, livid.   Requires fine air : 2° $n\,p\,\eta$, on a line pointing
to ε ; the further from η of two stars near together.

57—xi^h 22^m N 40° $6'$—6, 9—$5\cdot9''$—lucid white, violet.
9 variable ?

65. Triple—xi^h 48^m N 47° $15'$—7, $9\cdot5$, 7—$3\cdot8''$, $63''$—
bright white, pale purple, white.   Smyth suspects variation in
$9\cdot5$ and the distant 7, and that all may be connected.
2° $s\,f\,\chi$.

277 P XIII—xiii^h 54^m N 53° $47'$—$7\cdot5$, 12—$6\cdot8''$—
bright white, pale blue.   12 singularly distinct for its magnitude, as Smyth observes, and hence a fine test.   I saw it well.

[URSA MAJOR]

It lies in a string of stars reaching from ζ towards the coarse group in the hand of Boötes.

58 P X—x^h 17^m N 53° 20′—8, 8·5—3·6″—both white.

21—ix^h 16^m N 54° 37′—8, 9 — 6·3″ — silvery white, violet.

(290—xi^h 29·5^m N 46° 2′—6, 8—10″.)

(284—xi^h 28·9^m N 65° 17′—6·2, 7·8—2·1″—white and greyish. These two pairs, from Σ, I have not seen.)

Nebulæ.

81 and 82 M—ix^h.44^m N 69° 46′. Two nebulæ $\frac{1}{2}$° apart: 81 bright, with stellar nucleus, finely grouped with small stars, one of which is projected upon the haze. With a power of 80 I divided a little pair, 1386 Σ, $s\,p$—9, 9—2″. 82 is a curious narrow mottled ray.

43 ♅ V—xii^h 12^m N 48° 4′. Bright, oval, denser in centre.

97 M—xi^h 7^m N 55° 46′. Large pale planetary nebula: a very curious object. H gives it a diameter of 2′ 40″, which at the distance of 61 Cygni only, would fill a space equal to seven times the orbit of Neptune. He sees its light perfectly equable, with only a softened edge; but the more powerful instrument of the Earl of Rosse shews two large perforations and a resolvable spiral arrangement — a striking instance of the advantage of a larger aperture. 2° $s\,f\,\beta$. One of the very few omissions of the Map of S. D. U. K.

46 ♅ V—xi^h 3^m N 56° 25′. Elongated; a small star in the centre. 1° $s\,f\,\beta$.

205 ♅ I—ix^h 12^m N 53° 6′. Dull, in a fine field with 37, 6 mag. 1$\frac{1}{2}$° $s\,p\,\theta$. There are several pretty little pairs in the neighbourhood.

URSA MINOR.

This constellation is distinguished by a still more inappropriate length of tail than its larger neighbour, by which, as
Smyth observes, it is swung round every 24 hours: at its
extremity stands the most valuable star in the heavens —
Polaris — the first of the following list. — 1° 32′ from the
celestial pole, towards which, from the precession of the equinoxes, it will approach till A.D. 2095, when its distance will
be 26′ 30″, subsequently widening again.

Double Stars.

α—1ʰ 7ᵐ   N 88° 34′—2·5, 9·5—19″—topaz yellow, pale
white, 1830, 1838, 1849.   Dawes 9·5 bluish: and so I see it.
It is a common test, but only suited for small instruments, being
very easily seen with any aperture much exceeding 2 inches.
Dawes has proposed it as a general standard, finding that 2
inches and a power of 80 will shew it if the eye and telescope
are good.   In the Dorpat achromatic it has been seen by day.

π¹—Σ xvʰ 39·3ᵐ   N 81° 6′—6·1, 7—30″—both yellowish.
(Smyth yellow, blue.)   Not in Map of S. D. U. K., but easily
found from ε and ζ.

VIRGO.

A constellation especially remarkable, for those possessed of
adequate optical means, on account of the wonderful *nebulous
region*, in which a far greater number of these extraordinary
objects are accumulated, than in any other equal area of the
heavens; ♅ having detected within its boundaries no less

than 323. Few of them, however, are individually interesting; it is the mysterious thronging together of systems, each in itself an insulated aggregate of countless suns, that opens such a field for curiosity; nor would that curiosity be diminished if it should ever appear probable that the components of those systems are inferior to our own Sun in magnitude, and less remote from us than was supposed by ♅. They are in general so similar to each other, that I have only adduced a few as specimens. They are profusely scattered over this quarter of the sky; but the region more especially referred to is pretty well defined to the naked eye by the stars ε, δ, γ, η, and β ♍, and β ♌.

Double Stars.

γ—XII^h 35^m S 0° 41′—4, 4—7·5″, 1790, ♅; 1″, 1834; 0″, 1836; 2″, 1843; 3·5″, 1857—silvery white, pale yellow; white the brighter, but colours not always of the same intensity. Σ thought them alternately variable in brightness, with a possible period of at least several years. Fletcher white, yellow, colours fixed, mags. alternately variable, 4, 4·5. Period uncertain, owing to some anomalies; but under 200 years. Adams thinks 174 years most probable. As this wonderful pair has been widening ever since they closed up out of telescopic reach in 1836 (when, however, there was probably no actual *occultation*, as H ingeniously observes, since there was no decrease of brightness to the naked eye) a very moderate instrument will now shew them.

θ. Triple—XIII^h 3^m S 4° 47′—4·5, 9, 10—7·2″, 65″— pale white, violet, dusky.   H and South called 9 a very severe test for a 5 feet telescope, 1824.

R 2

[VIRGO]

84—XIII^h 36^m N 4° 15′—6, 9—3·5″—yellowish, smalt
blue. [difficult—a test.] Σ gives them common proper motion.

17—XII^h 15^m N 6° 5′—6, 9—20″—light rose tint, dusky
red. Σ as the last.

196 P XII—XII^h 44^m S 9° 35′—6·5, 9·5—33″—topaz
yellow, lucid purple. Closely *s p* ψ.

54—XIII^h 6^m S 18° 5′—7, 7·5—5·7″—both white, 1839
[pale yellow? pale blue? 1852].

238 P XIII—XIII^h 48^m S 7° 22′—7, 8·5—2·5″—both
white. Clearly divided with my 80.

32 and 33 P XII—XII^h 11^m S 3° 11′—7·5, 7·5—21″—
both silvery white. 3° *s* from η, a little *p*.

81—XIII^h 30^m S 7° 9′—7·5, 8—2·8″—bright white, yellowish. Binary?

25 P XIII—XIII^h 8^m S 10° 37—7·5, 8·5—42″—both
bluish. 2½° *p a*.

221 P XII—XII^h 48^m N 12° 15′—7·5, 9—29″—pale
white, sky-blue. 2° *p* ε, a little *n*.

127 P XIII—XIII^h 27^m N 0° 24′—8, 9—1·7″—pale
white, yellowish. I could only elongate doubtfully, 1851,
this interesting pair, which may probably have a period of
about 240 years. It lies closely *n p* ζ.

171 P XIII—XIII^h 36^m S 3° 34′—8, 10·5, 1830—30″—
light orange, pale lilac. [10·5 visible much out of focus, and
in strong moonlight, 1852.]

[About XII^h 24^m N 2° 30′ is a fine 8, 8·5 mag. pair.]

Nebulæ.

43 ♅ I—XII^h 33^m S 10° 50′. Elongated : in a beautiful
low-power field. A very fine and singular 7 mag. group lies *n p*.

[VIRGO]

60 M—XII^h 37^m N 12° 19'. In field with two others, 59 M, p, fainter, and H 1402, a minute object, with my 64 like a star, hazy with 80 ; lying between two small stars.

70 ♅ I—XIV^h 22^m S 5° 21'. Close to an 8 mag. star, and prettily grouped with smaller ones. H resolves it into 19 mag. stars. " So that here," as Smyth says, " we find another universe in the plenitude of space ! "

88 M—XII^h 25^m N 15° 12'. Elongated and dull. The nebulous district in which it lies is very wonderful ; 6 or 7 were swept over with 64, some of them tolerably conspicuous ; occasionally two in the same field : among them were probably 87, 89, 90, 91 M.

49 M—XII^h 23^m N 8° 46'. An inconsiderable object, but beautifully situated between two 6 mag. stars. A bright open pair lies s of it.

74 and 75 ♅ II—XII^h 46^m N 11° 59'. Two very faint objects in one field ; $n p$ (74) the larger and brighter : 75 is beautifully grouped with three stars. They are easily found, at a little distance p 221 P XII, which again is near ε.

VULPECULA.

A little modern asterism, in which its former, Hevelius, perceived a new star in 1672 ; but it continued visible for only two years, and has not been since identified. According to Chacornac, who has recorded many such changes, they are more frequent, even in modern times, than has till of late been suspected ; the improvement of catalogues enabling us to discriminate between the " mistaken entries " of the earlier observers, and real alterations.

Double Stars.

320 and 321 P XIX—xix^h 47^m N 19° 58'—7, 7—43"—
both white. Another pair p makes up a pretty group.

415 P XIX—xx^h 1^m N 20° 42'—8, 10—4·5"—pale white,
sky blue. In a glorious Galaxy field, 1° p θ Sag.

Nebula.

Some of my readers may perhaps feel that I have allotted an
undue proportion of space to minute and inconspicuous
objects. It may be so. I may have erred in supposing that
others might receive as much pleasure as myself from their
contemplation: yet a good many comprised in my original
scheme have been passed by, as well as a great mass of remarks
on the beauty or singularity of those which have been selected.
But, should I have failed in communicating to others a portion
of my own interest as to some parts of this list, it will be
closed with a nebula which I think will not be found disap-
pointing.

27 M—xix^h 54^m N 22° 20'. The "Dumb Bell" Nebula.
In a rich field we find two oval hazy masses in contact, of
which p seems to me the brighter, as it did to H. His
reflector failed to resolve it, but shewed the dark notches
filled in and made protuberant by faint luminosity, converting
the whole figure into an ellipse. The Earl of Rosse's 3 feet
speculum reaches its starry components: his 6 feet surrounds
it with an external ring having a neck like a retort. Bond's
achromatic also resolves it, but there the form shewn in small
instruments is lost. In gazing upon this glorious host of ten
thousand suns, in reflecting upon its awful distance, its incalculable magnitude, and the wonderful conditions of existence

where night would seem to be effaced amid the universal
splendour, the magnificent apostrophe of Kepler, which closes
his speculations on the habitability of our own Sun, may well
express our feelings, and form at the same time a most appropriate conclusion to the varied scenes which have passed in
review since we commenced these pages :—

"Abrumpo consultò et somnum et speculationem vastissi-
mam ; tantum illud exclamans cum Psalte Rege :

"Magnus Dominus noster, et magna virtus ejus, et sapientiæ
ejus non est numerus : laudate eum cœli, laudate eum Sol,
Luna, et Planetæ, quocunque sensu ad percipiendum, quâ-
cunque linguâ ad eloquendum Creatorem vestrum utamini :
laudate eum harmoniæ cælestes, laudate eum vos harmoniarum
detectarum arbitri : lauda et tu anima mea, Dominum
Creatorem tuum, quamdiu fuero : namque ex ipso et per
ipsum et in ipso sunt omnia, καὶ τὰ αἰσθητὰ, καὶ τὰ νοερὰ ; tam
ea quæ ignoramus penitus, quam ea quæ scimus, minima
illorum pars ; quia adhuc plus ultrà est. Ipsi laus, honor, et
gloria in sæcula sæculorum. AMEN."

THE END.

Printed in the United States
By Bookmasters